Table of Contents

Preface

Before you begin reading...

PART I

Introduction to Part I

Chapter 1 - Pythagoras
The greatest mystery in the history of mathematics. 1-1

Chapter 2 - Continuing with Pythagoras
The Pythagorean theorem has been celebrated as a basic truth in geometry for 24 centuries. But is this an illusion? 2-1

Chapter 3 - Are "irrational" numbers really irrational?
The discovery of "irrational" numbers caused the demise of the 500 year-old school of mathematics founded by Pythagoras. 3-1

Chapter 4 - Another conversation with Pythagoras
Was Pythagoras right all along? Was the universe created using natural numbers? 4-1

Chapter 5 - The Mathematical Rosetta Stone decodes "pi" and "e" 5-1

Chapter 6 - The Mathematical Rosetta Stone decodes Euler's Constant and the Golden Mean 6-1

Chapter 7 - The Mathematical Rosetta Stone
Dr. Einstein believes that "randomness" is an illusion. Perhaps we can prove him right! 7-1

Table of Contents

Chapter 8 - Dr. Einstein sets the record straight
1913 banner headlines across the world screamed in huge black print: "Einstein overturns Newton's laws of gravity." 8-1

Chapter 9 - René Descartes
Reveals how he made the startling discovery of a mysterious connection between Algebra and Geometry—a connection that mathematicians believed did not exist 9-1

Chapter 10 - Dr. Albert Einstein
Einstein's concept of space and Zeno's Paradox 10-1

Chapter 11 - Dr. Albert Einstein
What did Galileo, Newton, Maxwell, Faraday, and Einstein have in common? 11-1

Chapter 12 - Sir Isaac Newton
A new kind of mathematics, the secret practice of illegal "black arts," and Newton's fascination with the Bible 12-1

Chapter 13 - Dr. Richard P. Feynman, Nobel Laureate
A conversation about the strange mysteries of light, quantum mechanics, and Zeno's Paradox 13-1

Chapter 14 - Another conversation with Dr. Einstein
Do primes exist... or are they the greatest mathematical myth in history? 14-1

The Weird and Wonderful World of Mathematical Mysteries

by
James J. Asher
Prize-winning Writer

First Edition ISBN: 1-56018-507-4
Second Edition ISBN: 1-56018-038-2
Third Edition ISBN: 1-56018-075-7

Published by
Sky Oaks Productions, Inc.
P.O. Box 1102
Los Gatos, California 95031 USA
Phone: (408) 395-7600   Fax: (408) 395-8440
e-mail: **tprworld@aol.com**

To order other books by Dr. Asher, click on:
www.tpr-world.com

© Copyright 2001, 2004

All Rights Reserved Worldwide. No part of this publication may be reproduced, stored in a retrieval system, or transmitted in any form or by any means; electronic, magnetic tape, mechanical, photocopying, recording or otherwise, without permission in writing from an executive officer of Sky Oaks Productions, Inc.

This book is protected internationally by the Universal Copyright Convention of Geneva, Switzerland.

Table of Contents
PART II

Introduction to Part II

Chapter 15 - Pythagoras
A debate between the renowned Pythagoras and a team of amateur mathematicians, Farber and Heisel 15-1

Chapter 16 - Euclid
Euclid explains why his book of geometry, reprinted in a unbelievable 2,000 editions, is still a best selling book after 24 centuries 16-1

Chapter 17 - Pierre de Fermat
The Prince of Amateur Mathematicians finally reveals his marvelously simple solution to a mystery that has baffled mathematicians for hundreds of years 17-1

Chapter 18 - Decimals - real or just an illusion?
A conversation about decimals with the amateur mathematicians, Farber and Heisel 18-1

Chapter 19 - Christian Goldbach
The famous Goldbach Conjecture about arithmetic has a kindergarten simplicity that has baffled mathematicians for hundreds of years. Is it now solved? 19-1

Chapter 20 - Augustine Louis Cauchy
Negative numbers - the greatest "red herring" in the history of mathematics?. 20-1

The Players
Brief biographical sketches of Pythagoras and the other mathematicians in this book Players-1

Selected References References-1

PREFACE

This book started with a 2,000 year old mystery about something as simple as a ratio. Everyone is familiar with ratios such as 1/2, 1/3, 1/4 and so on. So, what is the mystery? 2,000 years ago, one of the most quoted mathematicians, Pythagoras, discovered that *harmony* from musical instruments such as the lute is determined by the ratio of the length of the strings.

Pythagoras observed that the planets too seem to orbit in harmony to ratios. Ratios, he reasoned, must be the mystical concept the Divine Mathematician used to create the universe and everything in it. For 500 years his school of mathematics flourished until a "flaw" was uncovered in the sacred concept of the ratio.

A "perfect" ratio such as 1/2 or 1/3 or 1/4 had to be composed of natural numbers. But, the square root of 2 is a problem. It is not a ratio at all, but a non-repeating decimal number that continues on into infinity! How can this be? Pythagoras thought, "This goes against the Divine Blueprint for the universe. It is imperfect - too messy. It can't be... and yet there it is." This disturbing non-ratio became known as the infamous, "irrational" number. Soon after this, mathematicians began discovering "irrational" numbers everywhere they looked, including pi, e, and the Golden Mean.

There are two principles that guide both mathematicians and scientists. One is the principle of *parsimony* (the simple explanation is preferred to the complex) and the other is *symmetry*. If Pythagoras was right, then there is no such thing as "irrational" numbers. This has to be an illusion, a red herring that has misguided mathematicians into a "blind alley" for centuries.

The simple explanation is that every number, no matter how complex it appears, has an underlying ratio comprised of whole numbers. I know about Euclid's logic "proving" that the square root of 2 cannot—absolutely cannot—be transformed into a ratio of whole numbers. Although Euclid was right in his original conclusion, there is an exciting solution that he never considered—*now published for the first time in Chapter 4...* It is called The Mathematical Rosetta Stone, which one reviewer called "an ingenious insight" into a 2,000 year old mystery about irrational numbers.

To make this book an easy read, I decided to use the medium of conversation. After all, Plato was successful in explaining geometry to a slave boy using his famous "dialogues."

The idea of using conversations with mathematical celebrities came to me as I watched an actor playing Jack London, the first American writer to earn a million dollars in royalties. The actor looked and talked just like Jack London as he told us about events in his life—and then turned to the audience for questions. One person asked: *"Did you experience any rejections before you sold your first story?"* Mr. London answered: *"I had 643 rejections before selling my first short story. After I was a successful writer, every once in a while I would slip in one of those 'rejected' stories and sure enough, it was rejected by the editor every time I tried."*

I was fascinated by the process of becoming so intimate with the details of a person's life that questions could be answered spontaneously. As I moved deeper and deeper into the lives of Pythagoras, Newton, Descartes, Einstein and the others, I was surprised at the answers I got from these celebrities.

Meanwhile, I invite you to enjoy "after-dinner" conversations with some of the greatest minds in history.

Before you begin reading, let me make a few comments...

Dr. Watson played an important role in the Sherlock Holmes mysteries by asking questions that motivated the master sleuth to consider unlikely suspects that might otherwise go undetected. Jon Robertson was my Dr. Watson, asking questions about puzzling mathematical relationships that bother most people, except mathematicians. For example, if **2** times **2** = **4**, why doesn't **-2** times **-2** = **-4**?

This is an out-of-the-box look at some exciting mathematical mysteries. 4,000 studies of brain lateralization suggest that most people can understand anything if the presentation plays to their right brain. The problem is that in school, mathematics is a foreign language that plays to the left brain with an obscure code—an alien language that makes understanding impossible for most people. If you are one of them, get ready to understand exquisite mathematical relationships that you imagined were "not your cup of tea."

Take a bow

Dr. Warren Esty for your patience in answering my inquiries about mathematical proofs—and every time you answered, you added the charming tag line, *"Ask me another question."*

Dr. Kenneth Bradshaw for answering my many "odd ball" questions. (Dr. Bradshaw read a draft and offered scores of helpful suggestions.)

Virginia Lee Asher for your careful line-by-line proofing and probing questions that inspired me to clarify many obscure issues during the writing of this book.

Part I

Part I is exciting because <u>for the first time in print</u>, I reveal a new discovery that solves a 2,000 year old mystery—something that baffled famous people such as Pythagoras, René Descartes, Albert Einstein, Sir Isaac Newton, and Richard P. Feynman.

Pythagoras is amazed when I demonstrate to him that the so-called "irrational" numbers have a hidden pattern which makes them rational. He believed this was true all along, but could not prove it. The debunking of irrational numbers delights Dr. Einstein because it is further evidence that the universe is not the product of random, unpredictable events. "The Old One," as Dr. E. liked to say, "did not throw dice to create the universe."

René Descartes' belief in a precise, clockwork universe inspired him to search for and finally discover the hidden connection between geometry and algebra—the startling finding that is fundamental to innovations in modern science and technology. More patterns from Sir Isaac Newton predict with mathematical precision the attraction of planets separated by millions of miles.

Then Dr. Richard P. Feynman explores the mystery of light and reluctantly returns to "randomness" to explain the behavior of individual photons. However, "I believe," as Dr. Einstein would say, "this is only temporary until we discover the true cause- effect relationship."

Part 1

CHAPTER 1
The greatest mystery in the history of mathematics:

A Conversation with Pythagoras*

JJA: Pythagoras, you are said to be one of the three greatest mathematicians of all time. It is an honor to talk with you.

Pythagoras: Thank you.

JJA: Your famous school of mathematics was founded in about 530 B.C. and continued in operation for an astonishing 500 years. How do you account for that—especially given the secretive nature of your organization?

Pythagoras: I think that we understood how "The One" created the universe and everything in it. There was a truth that people responded to.

JJA: The One?

Pythagoras: "The Good"—the agent or intelligence that created the universe.

JJA: You believe that mathematics is the key to reality.

Pythagoras: Absolutely. "The One" is a Mathematician.

---

*In the back of this book in a section called **The Players**, you will find brief biographical sketches of Pythagoras and other mathematicians.

JJA: You are quoted as saying, "Number is everything!"

Pythagoras: I believed that nature can be explained with whole numbers and especially ratios of whole numbers.

JJA: "Ratio" seems to be a sacred concept to the Pythagoreans.

Pythagoras: It is. I believed that everything from harmony in music to the movement of the planets is produced by certain ratios.

JJA: Scientific discoveries appearing hundreds of years after you, confirm that whole numbers are in equations that access us to predictions and often control of atoms, electricity, gases, light and energy.

Pythagoras: Yes. I noticed that Kepler, Newton, and Einstein all discovered whole numbers and ratios of whole numbers in their famous mathematical equations.

JJA: All of those scientific celebrities consider you a pioneer in thinking about nature and the universe. Dr. Einstein was a big fan of yours. He also believed that the planets revolve to the metronome of a pre-ordained clock—wound by the Divine Clockmaker of the universe. He was fond of saying, as I recall: "God is not uncertain. God is certain. If God was uncertain, science would not be possible."

Pythagoras: Dr. Einstein, like myself, was aggravated with any suggestion that the universe came to be by chance, by a random coalition of uncertain events or by mysterious gases that explode with a "big bang."

JJA: You were several thousand years ahead of your time. But, suddenly and unexpectedly, your school lost credibility. What happened?

Pythagoras: It had to do with my theorem.

JJA: The famous Pythagorean theorem about right-angle triangles, which looks like this:

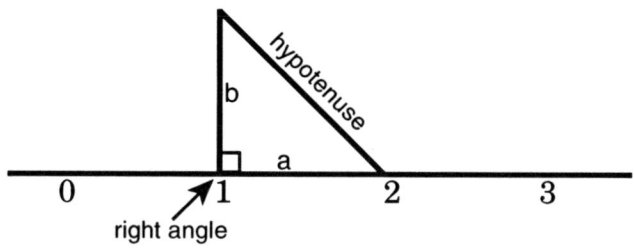

Pythagoras: Because the hypotenuse is on a *slant*, it cannot be measured directly. My theorem provided an indirect way to measure that *slanted line.*

JJA: Why is this important?

Pythagoras: Because it has so many applications. It was used to construct the great pyramids of Egypt. It was used in architecture, in surveying, in estimating the distances between planets...

JJA: How does it work?

Pythagoras: The goal is to *somehow* measure the slanted line, which is the hypotenuse. We can't always measure it directly.

JJA: Why not just put a ruler on the slanted line?

Pythagoras: Ok. It will work for the picture above because it's on a small piece of paper, but what if the slanted line is very large or goes through the air, like to the top of a mountain?

JJA: You discovered a way to measure the slanted line indirectly, right?

Pythagoras: Yes.

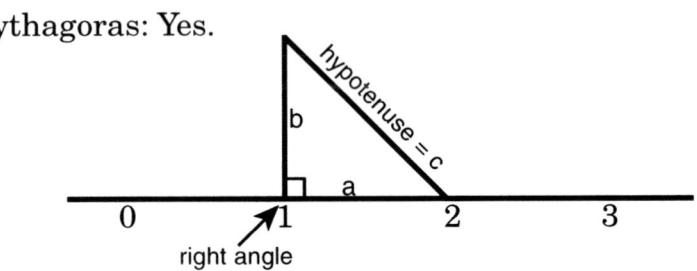

First, $a^2 + b^2$ will equal c^2.

Second, $\sqrt{a^2 + b^2}$ will equal c (the actual length of the slanted line).

JJA: If your theorem is so valuable, why did your marvelous school of mathematics fade away?

Pythagoras: I'm almost ashamed to tell you what really happened. It has to do with a right-angle triangle with one leg measuring 1 and the other leg also measuring 1.

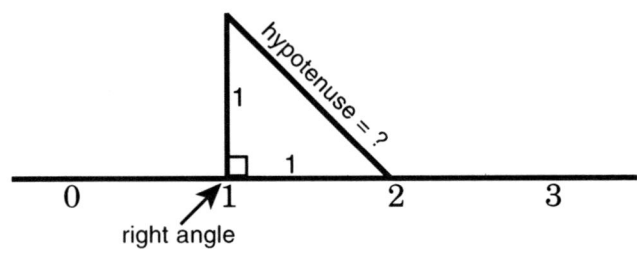

JJA: Let's see... applying your theorem,

First, $1^2 + 1^2$ will equal **2**.

Second, $\sqrt{1^2 + 1^2}$ will equal $\sqrt{2}$ (the actual length of the slanted line).

So the length of the hypotenuse will be $\sqrt{2}$. What's wrong with that?

Pythagoras: To understand, we should explain the nature of square roots.

JJA: Which is?

Pythagoras: Which is revealed when you decode a square root. For example, how would you decode the square root of 4, which looks like this: $\sqrt{4}$?

JJA: The answer is 2 because 2 times 2 equals 4.

Pythagoras: How do you decode $\sqrt{16}$?

JJA: The answer is 4 because 4 times 4 equals 16.

Pythagoras: Good! Now try decoding $\sqrt{2}$. What two <u>identical</u> <u>numbers</u> can you multiply together to equal 2?

JJA: None. There is no number multiplied by itself that will equal 2.

Pythagoras: Precisely. This is a messy situation because I believe that "The One" is a tidy thinker. Everything in nature should fit together perfectly.

JJA: You discovered an important relationship that was not neat and tidy.

Pythagoras: Yes.

JJA: You were devastated.

Pythagoras: To say the least. I asked everyone in the organization to calculate possible ratios in an attempt to find a solution. Nothing worked.

JJA: I understand that you made members of your organization take an oath of silence.

Pythagoras: Yes, but of course, it didn't work. Eventually the secret "leaked."

JJA: Later generations of mathematicians would call $\sqrt{2}$ an "irrational" number because it cannot be converted into a ratio of whole numbers. For comparison, $\sqrt{4}$ is "rational" because it can be expressed as the ratio of $\frac{2}{1}$.
After that, mathematicians "discovered" scores of so-called "irrationals," the most famous ones being:

pi = 3.14159265358979323846...

e = 2.718281828459045...

the **Golden Mean** = 1.618033988749894...

and **Euler's Constant** = .577215664901532...

Pythagoras: It is maddening that the integers after the decimal in those "irrationals" go on and on without a pattern and seemingly forever as shown by three dots. It is contrary to everything I believe. It is messy! It does not fit with my concept of a universe that runs with the precision of a clock.

JJA: With a powerful computer, do you think you could have found that illusive ratio for $\sqrt{2}$?

Pythagoras: You picture the ratio hiding like a needle in a mathematical haystack of several trillion calculations?

JJA: Possibly.

Pythagoras: Your modern-day mathematicians would say, "Absolutely not!" I believe Euclid proved that this quest to find "the ratio" for $\sqrt{2}$ is impossible.

JJA: What if I told you that Euclid was right— there is no (one) ratio of whole numbers that will generate the $\sqrt{2}$. But, it never occurred to Euclid that there could be a set of ratios that will generate the $\sqrt{2}$ and all other so-called "irrational" numbers. I believe I have found the key to these hidden ratios.

Pythagoras: I would be thrilled! What do you call this key?

JJA: I call it, the "Mathematical Rosetta Stone."

Pythagoras: Will it explain other notorious "irrationals" such as pi, e, the Golden Mean and Euler's constant?

JJA: Yes.

Pythagoras: When will you reveal this. I must know how it works!

JJA: It is coming up in a few chapters. But next, your famous theorem is being challenged by two amateur mathematicians. Are you up to a debate?

Pythagoras: I believe I can hold my own in any debate.

CHAPTER 2

The Pythagorean theorem has been celebrated as a basic truth in geometry for 24 centuries. But is this an illusion?

A discussion with Pythagoras

JJA: I have to ask you the most intriguing question first. Why did you hide behind a curtain so that your students only heard your voice? Then at graduation, you stepped from behind the curtain and made your appearance, bowing from the waist, I presume.

Pythagoras: Didn't you ever hear of showmanship?

JJA: In other words, this was not some personality aberration. This was, to use a contemporary phrase, show business.

Pythagoras: Certainly. I generated unbearable curiosity in my students. Their reward for good work? They got to see the Maestro in person.

JJA: So this was not a kind of interpersonal problem that, Sigmund Freud, for example, experienced.

Pythagoras: I have not followed the originator of psychoanalysis. What was his personal problem?

JJA: He was uncomfortable facing his patients when they were describing intimate details of their lives. So, he asked the patient to lie on a couch facing away from him while he sat in a chair behind the person's head and scribbled notes.

Pythagoras: Fascinating. But, I assure you that facing my students was not difficult for me. My intent was to keep them focused, attentive, and keenly interested in the mysterious figure behind the curtain.

JJA: It occurs to me that God worked in a very similar fashion.

Pythagoras: How so?

JJA: In our old testament there are stories of ancient people hearing the voice of God speaking to them but no one ever actually saw God.

Pythagoras: Then I have an excellent model for my theatrical performance.

JJA: I should say so. Now to your work. We often read that the ancient Greek mathematicians proved their theorems with diagrams, rulers and a compass, but no algebra.

Pythagoras: Right. Algebra was probably discovered in 1700 B.C. by an Egyptian mathematician named Ahmes. But, the "father of algebra" is often recognized as Diophantus, who developed the symbols and implications. He lived in 200 A.D., about 800 years after me.

JJA: I recall that the most dramatic improvements in algebra did not appear until after the middle ages, in 16th century Europe.

Pythagoras: Yes. European scholars discovered the intriguing symbols and structure of the algebraic language and began to explore its mysteries.

JJA: That being the case I don't understand. You, Euclid and others offered formulas such as your famous theorem of $a^2 + b^2 = c^2$, referring to a right-angle triangle. Isn't that algebra?

Pythagoras: Well, not exactly. We understood how to add, subtract, multiply and divide. We understood that 2 added to 2 equaled 4. We were, of course, limited to positive integers only— that is, we were limited to the natural numbers, sometimes called the counting numbers of 1, 2, 3 and so forth. Negative numbers did not appear until the middle of the 16th century.

JJA: But isn't 2 + 2 = 4 an equation and isn't an equation the biggest jewel in the crown of algebra?

Pythagoras: This is tricky. Although we did comprehend that if you added two of something to two of something, the result is four of something, we did not know that this was an "equation."

JJA: Forgive me. I don't mean any disrespect. But, so what?

Pythagoras: The point is that an "equation" is a concept which has implications. It is kind of a trump card in the game of algebra. The game has rules for playing which we did not know. Hence, we could not play the game. We were limited to something and something equals something else.

JJA: It's still fuzzy to me.

Pythagoras: In your century or perhaps the century before you, it was demonstrated that two parts of hydrogen and one part of oxygen would produce water. Although this is an "equation," it is not algebra. All we have is a compound of elements that result in another element which is water. It is addition only.

JJA: How does this apply to your famous theorem?

Pythagoras: For example, take my theorem of $a^2 + b^2 = c^2$. Let's say that the bottom side is 3, and the vertical side is 4. My theorem predicts that $3^2 + 4^2 = 5^2$

JJA: ...and 5^2 should be the hypotenuse.

Pythagoras: Yes. Let's see if I can prove that this is true.

Let's start by drawing a right-angle triangle here in the sand.

Next, if the bottom side of the triangle is 3, then the square of 3 is 9 and looks like this—

JJA: Ok. Now, you will draw another square on the vertical leg of the triangle. What is the length of the vertical leg?

Pythagoras: The length is 4. So let's draw the square of 4 which is 16.

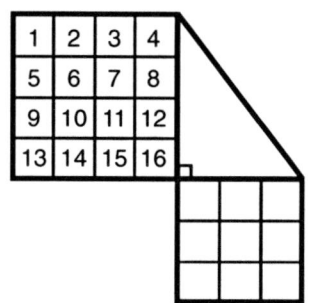

Pythagoras: Now when you add the two squares together what do you get?

JJA: 9 square inches and 16 square inches is 25 square inches.

Pythagoras: 25 square inches is five squared which when drawn looks like this—

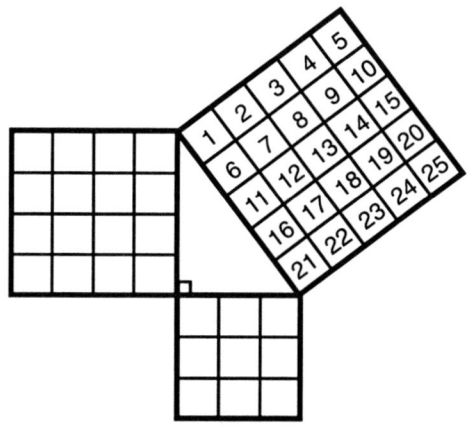

JJA: Voila! You have a demonstration that when we add the square of the two legs of a right-angle triangle, the result is the square of the hypotenuse.

Pythagoras: Precisely. But notice that the demonstration is with diagrams, not algebra.

JJA: Now for a silly question.

Pythagoras: A silly question?

JJA: Yes. So what? Why is this such a famous theorem? Who cares?

Pythagoras: What made this discovery exciting is this: The relationship of the sides of a right-angle triangle to the hypotenuse was probably know in Babylon 1000 years before I was born in the year 580 B.C. My contribution was in applying deductive logic to show that the relationship was valid. The process is one of starting with a premise and following a chain of reasoning to a conclusion.

JJA: I noticed that your theorem inspired more deductive "proofs" through the centuries than any other idea in geometry. Even one of our American presidents, James Abram Garfield, in 1876 discovered a proof of your theorem using the concept of a trapezoid. It was published in the *New England Journal of Education.*

Pythagoras: Yes, I saw some flaws in the proof, but we will save that for another interview.

JJA: Clearly, your contribution has entertained thousands of math buffs.

Pythagoras: Not only entertained, but the concept has been used in astronomy, architecture, and construction—including the building of the great Egyptian pyramids in which there is no flaw so large that you cannot cover it with your thumb.

JJA: Amazing. Tell me, how did they cut and lift those monstrous stone blocks 40 stories high and fit them together with such precision?

Pythagoras: That is a secret only the Egyptians know. Ask me something that I am familiar with.

JJA: You and your colleagues did a splendid job with measuring straight lines. How about curved lines?

Pythagoras: Don't think that we did not play with curved lines to discover how to measure them. We knew how to measure straight lines so we could find, for example, the area of a square, the area of a rectangle or the area of a triangle.

JJA: And curved lines?

Pythagoras: Curved lines are another measurement mystery. It would not be until the 17th century when Newton in England and Liebnitz in Germany discover an approach to measuring curves which you know as calculus.

JJA: Why do you say, "approach"? This sounds so indefinite?

Pythagoras: It is indefinite. It is an approximation because no one yet has a precise measurement for curved lines, but why not ask Sir Isaac Newton about this?

JJA: I will. Any recommendations on how to interview the great Sir Isaac Newton? I understand he is quite sensitive and seclusive.

Pythagoras: He has good reason to be. He expected elation when he first presented his ideas about curved lines to the Royal Society of Science in London, but the audience looked like they were attending a funeral. They were cold and silent. These strange concepts did not fit anywhere in the mathematical reality of 17th century scientists. The ideas simply did not make sense. They dismissed Newton's presentation as "bloody nonsense."

JJA: But, Newton had a reputation for "good science."

Pythagoras: Yes, for example, the scientific belief at that time was that white light was pure and the colors were impure. Newton demonstrated with his famous prism that the reverse is true. White light is impure and the colors are pure.

JJA: Didn't his reputation as a scientist of the first rank give him credibility with the audience?

Pythagoras: Yes, but they felt that he must now be psychotic with this mathematical scheme about curved lines.

JJA: What happened next?

Pythagoras: Let him tell you.

JJA: Before I do that, I want to ask you about two amateur mathematicians who lived circa 1930 by the name of Theodore Farber and Theodore Heisel.

Pythagoras: Should I know them?

JJA: Yes, because they wrote a book that makes a rather simple but audacious statement which is—the Pythagorean theorem that has been celebrated for 24 centuries as a universal truth in geometry—that "truth" is a fiction.

Pythagoras: Tell me more about these gentlemen. And explain why I should give them any credibility.

JJA: Well, Theodore Heisel was so convinced that he and Farber had discovered some "universal truths" that professional mathematicians had overlooked, he self-published five thousand copies of his book and distributed them free to public libraries throughout the United States.

Pythagoras: Anyone with that commitment and sincerity deserves careful consideration. Specifically, what did they do?

JJA: Specifically, these two amateur mathematicians had the audacity to wrestle not one, not two but three respected gorillas that have roamed the mathematical jungle for centuries.

Pythagoras: Hmmm. Gorillas roaming the mathematical jungle. I like the metaphor. Tell me about them.

JJA: One is the Pythagorean theorem, the second is the value of decimals and the third is the validity of negative numbers.

JJA: I would like to get you together with them for a debate about your famous theorem.*

Pythagoras: I'm up to the challenge!

Summary

How Pythagoras proved his famous theorem without using algebra. He pointed out that his theorem was used to construct the great Egyptian pyramids. He comments on the mystery of measuring "curved lines."

---

* The exciting debate between Pythagoras and the two amateur mathematicians, Farber and Heisel, is in Part 2 of this book.

CHAPTER 3

The discovery of "irrational" numbers caused the demise of the 500 year-old school of mathematics founded by Pythagoras.

Are "irrational" numbers really irrational?

JJA: As I have stated many times, there is no one in the history of mathematics who is quoted more often than you.

Pythagoras: Thank you.

JJA: I believe that after many centuries, we can say with confidence that you were exactly on target with your belief that the laws of nature can be predicted and often controlled by whole numbers or ratios of whole numbers.

Pythagoras: It seems to be true of electricity, heat, light, temperature, and gases. Even Albert's equation of $e = mc^2$ involves whole numbers.

JJA: You are still convinced that the universe was created, and is regulated, by integers.

Pythagoras: Yes. Even Albert agrees. He is amused with complex mathematical models that mathematicians like to call *elegant*. "Elegance," he liked to say, "is for tailors. Nature obeys simple laws."

JJA: Today we are exploring the nature of "irrational" numbers.

Pythagoras: The amateur mathematicians make a persuasive case that "irrational" numbers do not exist because decimals do not exist. Of course, decimals exist but they are an artifact, an aberration—an abnormality of mathematics.

JJA: Do you think that decimals are unnecessary in calculation?

Pythagoras: Decimals distract from exact measurement. Decimals give us approximations whereas calculation with fractions will give us precise measurements.

JJA: You look as if you're still mulling that over.

Pythagoras: Yep, and I am struck by another strange relationship between fractions and decimals...

JJA: What's that?

Pythagoras: You can transform <u>all</u> fractions into decimals, but you can only transform <u>some</u> decimals back into fractions. That asymmetry bothers me.

JJA: According to contemporary wisdom, the decimals that you cannot convert back into fractions are the "irrationals." The fact that they cannot be converted makes them "irrational."

Pythagoras: I understand you don't believe this is a fact.

JJA: No I don't. I believe that <u>all</u> decimals can be converted into fractions. <u>I</u> <u>believe</u> <u>there</u> <u>are</u> <u>no</u> <u>irrational</u> <u>numbers</u>!

Pythagoras: You are a candidate for my brotherhood (which, incidently, included women). We also believed that all numbers were rational—until we encountered the infamous square root of 2! But you know that story.

JJA: Yes, indeed.

Pythagoras: Is your concept related to the Greek spiral showing that "irrationals" can be represented by finite, straight lines?

JJA: Refresh me on how this works. Can you draw something to show what we are talking about?

Pythagoras: First, you recall the problem with $\sqrt{2}$?

JJA: If we convert $\sqrt{2}$ into decimals, we get 1.414221356...

Pythagoras: Yes, the decimal equivalent of $\sqrt{2}$ begins with the integer 1, and then the numbers after the decimal continue flowing on without end.

JJA: And the Greeks showed us—what?

Pythagoras: The Greeks surprised us by demonstrating that if you draw a graphic representation of $\sqrt{2}$, there is a <u>definite beginning</u> and a <u>definite end</u>.

JJA: So how did they do it?

Pythagoras: Let me show you. I'll draw a horizontal line right here to represent the number line and mark it off at intervals of 0, 1, 2, and 3.

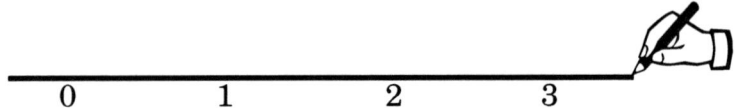

JJA: Now what?

Pythagoras: Now, I'll draw a vertical line at 1 on the number line, rising exactly one unit in height.

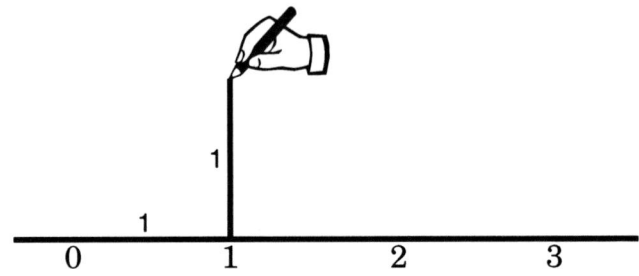

JJA: I can guess what you're going to do next...

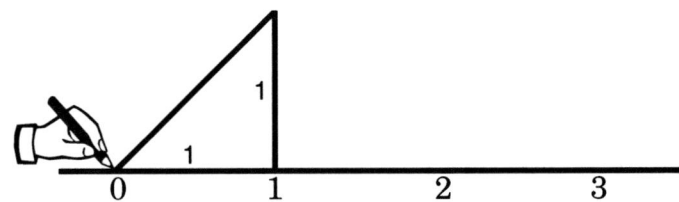

Pythagoras: As you can see, we now have what is known as a right triangle

JJA: And, using your famous theorem, the length of the hypotenuse will be equal to $\sqrt{2}$.

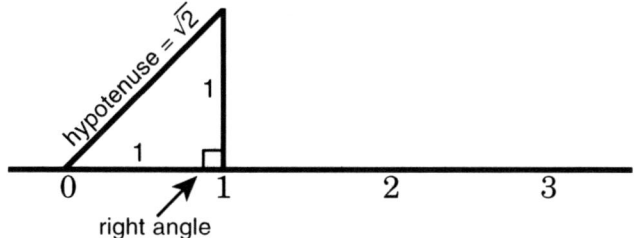

Pythagoras: If we can rotate the slanted line (the hypotenuse) downward until it's laying flat on the number line, we can measure its exact length.

JJA: I presume you'll use a compass to do this, right?

Pythagoras: Yes, I'll draw an arc like this...

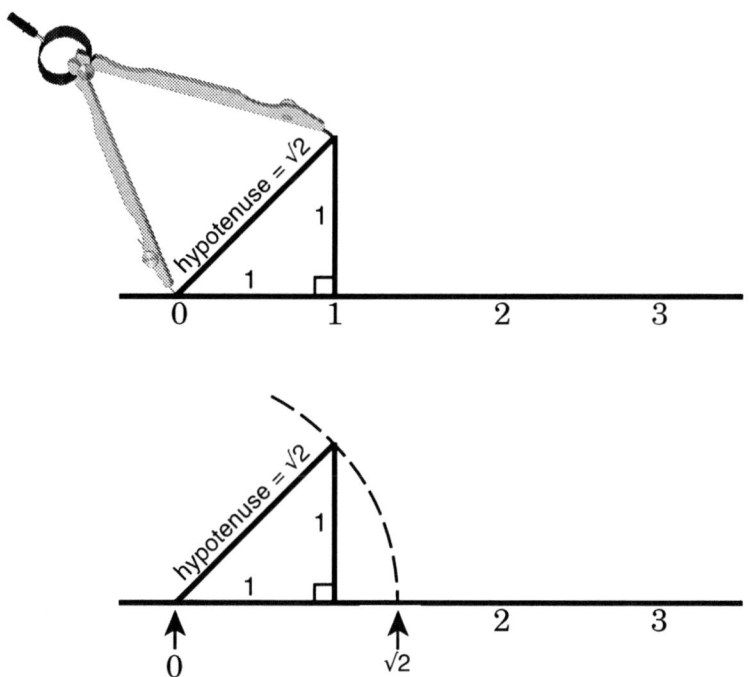

Pythagoras: Notice that the hypotenuse has a definite length on the number line, about half way between the 1 and the 2.

JJA: Wonderful. So you have demonstrated that the square root of 2 has a finite length and is certainly not infinite.

Pythagoras: Precisely! And using the same strategy we can find a finite length for the square root of 3, the square root of 4 and so forth.

JJA: Again, the conversion into decimals is a red herring. It misleads us into thinking that we have an actual number that increases into infinity.

Pythagoras: I wish I'd known this 24 centuries ago!

Summary

Pythagoras demonstrates that so-called irrational numbers such as square root of 2 and square root of 3 are located on the number line. They are finite with a definite beginning and a definite end on the number line.

The mystery is: When we convert numbers into decimals such as square root of 2 and square root of 3, there is no definite location on the number line. The decimal representations seem to expand into infinity.

CHAPTER 4
Was Pythagoras right all along?
Was the universe created using natural numbers?
Another conversation with Pythagoras.

JJA: Well, we circle around and come right back to the maestro. Were you and your secret mathematical brotherhood sandbagged with the discovery of the "irrational" number—the infamous square root of 2?

Pythagoras: Absolutely! If the universe can be explained with integers and ratios, then "irrational" numbers cannot exist, but there it is.

JJA: Let's see... applying your theorem, if one leg of a right angle triangle is 1 and the other leg is 1, then the hypotenuse will be the square root of 2.

Pythagoras: The square root of 2 almost drove me crazy. We could not find an integer or a ratio that converted into the square root of 2. My entire model of the universe was threatened.

JJA: What if I told you that your model of the universe is not threatened? What if I told you that your vision was accurate. *The universe can be explained with whole numbers!*

Pythagoras: You mean, that "irrational" numbers may be another mathematical red herring?

JJA: Yes, yes.

Pythagoras: Hooray! I knew it. I knew it! I just couldn't prove it. Tell me how it works...

JJA: I call the procedure for decoding an "irrational" number into a ratio, the Mathematical Rosetta Stone.

Pythagoras: I am confused. Didn't Euclid "prove" that it was impossible to decode an "irrational" number into a ratio?

JJA: Yes, he did and he is quite correct. It is impossible.

Pythagoras: I don't get it. Then what are you saying?

JJA: While it is impossible (although I don't like that word in either mathematics or science) to decode an "irrational" number into one ratio, it is not impossible to decode an "irrational" number into <u>more than one ratio</u>.

Pythagoras: Please, show me how it works.

JJA: All right. Let's start with your nemesis, the square root of 2. Notice that if we convert into decimals, we get a runaway series of seemingly random numbers that have no pattern—like this: $\sqrt{2}$ = 1.41421 35623 73095 04880 16887 24209 69807 85696 71875 37695...

I discovered how to decode the patternless series of numbers with my Mathematical Rosetta Stone—but before I share the secret with you, let me show you the results for $\sqrt{2}$ ...

On the left hand side of the chart below, you will see a unique ratio that corresponds to every integer in the $\sqrt{2}$, both before and after the decimal point.

Set of Ratios	$\sqrt{2}$ =	1.41	42	13	56	23	73	09	5
16 ÷ 11 =		1.4							
157 ÷ 111 =		1.41							
1571 ÷ 1111 =		1.41	4						
15714 ÷ 11111 =		1.41	42						
157135 ÷ 111111 =		1.41	42	1					
1571348 ÷ 1111111 =		1.41	42	13					
15713484 ÷ 11111111 =		1.41	42	13	5				
157134840 ÷ 111111111 =		1.41	42	13	56				
1571348403 ÷ 1111111111 =		1.41	42	13	56	2			
15713484026 ÷ 11111111111 =		1.41	42	13	56	23			
157134840264 ÷ 111111111111 =		1.41	42	13	56	23	7		
1571348402637 ÷ 1111111111111 =		1.41	42	13	56	23	73		
15713484026367 ÷ 11111111111111 =		1.41	42	13	56	23	73	0	
157134840263677 ÷ 111111111111111 =		1.41	42	13	56	23	73	09	
1571348402636772 ÷ 1111111111111111 =		1.41	42	13	56	23	73	09	5

JJA: With QBasic on my PC, my limit is 16 digits, but others with more powerful computers can easily continue the pattern into infinity.

Pythagoras: Amazing! Now, how did you happen to select the number one for the denominator?

JJA: You wouldn't believe the number of notebooks I have filled up with various ratios to discover a general pattern that would account for every single decimal in the so-called "irrational" numbers.

Pythagoras: Six hundred members of my brotherhood, together with myself, tried thousands of calculations looking for that illusive ratio representing $\sqrt{2}$—but all attempts failed. Everything I believed all my life to be true suddenly appeared to be false. I seriously thought about taking my own life.

JJA: I'm glad you didn't. I also tried hundreds of variations that didn't work.

Pythagoras: What kept you looking? Why didn't you give up?

JJA: I didn't give up because I simply believed that you were on the right track. Ratios had to exist—they just had to exist to explain those crazy "irrational" numbers. So I kept playing with the problem until something clicked.

Pythagoras: I can imagine the unbelievable excitement you felt when the hidden ratios appeared.

JJA: It was wildly exciting—knowing that I had discovered something that Euclid "proved" did not exist.

Pythagoras: As you mentioned before, Euclid was quite correct in that there is no single ratio that will represent $\sqrt{2}$. It did not occur to Euclid that the answer would be a set of ratios—instead of just one.

JJA: Our first task is to find a set of digits divided by or into 11 that will account for the first (and perhaps an additional digit) in the series. After the first entry with 11 or its expansion in either the numerator or denominator, the Mathematical Rosetta Stone (MRS) will decode an appropriate ratio for each successive digit in the "irrational."

Pythagoras: But I am curious. What happens when you divide by a different integer, let's say a 2 instead of a 1?

JJA: Excellent suggestion. Let's try it:

Set of Ratios		√2 = 1.41 42 13 56 23 73 09 5
31 ÷ 22		= 1.4
314 ÷ 222		= 1.41
3144 ÷ 2222		= 1.41 4
31427 ÷ 22222		= 1.41 42
314270 ÷ 222222		= 1.41 42 1
3142696 ÷ 2222222		= 1.41 42 13
31426967 ÷ 22222222		= 1.41 42 13 5
314269680 ÷ 222222222		= 1.41 42 13 56
3142696805 ÷ 2222222222		= 1.41 42 13 56 2
31426968052 ÷ 22222222222		= 1.41 42 13 56 23
314269680527 ÷ 222222222222		= 1.41 42 13 56 23 7
3142696805274 ÷ 2222222222222		= 1.41 42 13 56 23 73
31426968052734 ÷ 22222222222222		= 1.41 42 13 56 23 73 0
314269680527353 ÷ 222222222222222		= 1.41 42 13 56 23 73 09
3142696805273544 ÷ 2222222222222222		= 1.41 42 13 56 23 73 09 5

JJA: My conjecture is that the expansion of any whole number in the numerator or denominator will generate a ratio for each digit in the "irrational" number.

Pythagoras: Your demonstration verifies the conjecture. How about some of the other famous "irrationals" such as **pi** and **e** and the **golden mean**?

JJA: I'm glad you asked. Let's apply the Mathematical Rosetta Stone and decode each of those celebrated constants in the next chapter.

Summary

Was Pythagoras right all along? Was the universe created using natural numbers? Pythagoras looks at my never-before-published decoding strategy called the Mathematical Rosetta Stone, which explains every integer before and after the decimal in the "irrational" numbers.

Conclusion: All numbers, rational and irrational, can be converted into ratios!

CHAPTER 5
The Mathematical Rosetta Stone decodes the constants "pi" and "e"

JJA: Some background about **pi** may be helpful. Will you start us off.

Pythagoras: Certainly; **pi** is the circumference of a circle divided by the diameter and is the so-called "irrational" number of 3.14159265358979323846... **Pi** has entertained mathematical buffs for several thousand years.

JJA: The challenge is to find a ratio of whole numbers that will generate <u>every</u> digit <u>before</u> and <u>after</u> the decimal.

Pythagoras: China's Tsu Ch'ung Chi, who lived from 430 to 501 A.D., used the ratio of 355/113 to account for the following seven digits in **pi**: 3.141592.

JJA: If only that Chinese mathematician selected a "1" instead of a "3" in the denominator, the mystery of "irrational" numbers may have been resolved 1,500 years ago!

Pythagoras: I notice that interest in **pi** is still passionate, which is evident in a modern website entitled *"pi through the ages"* and lists the record number of digits calculated—more than 50 billion.

JJA: The largest number of decimal digits calculated so far is 51,539,600,000, a record held by Yasumasa

Kanada and Daisuke Takahashi from the University of Tokyo. The strategy for finding these digits in **pi** is based upon one of three mathematical strategies, none of which involve whole number ratios.

For example, Gregory proposed an infinite series in 1670. **Pi** = 4(1 -1/3 + 1/5 -1/7...) Strangely, if you continue working a million fractions in the series, you will only account for the first five decimal places in **pi**, which are: 3.14159...

Pythagoras: Now, your intent is to use your Mathematical Rosetta Stone to find a ratio that corresponds to each digit, both <u>before</u> and <u>after</u> the decimal in **pi**, correct?

JJA: Yes. So, **pi** = 3.1415926535897... and here we go...

Set of Ratios	**pi** = 3.14	15	92	65	35	89	7...
35 ÷ 11	= 3.1						
349 ÷ 111	= 3.14						
3490 ÷ 1111	= 3.14	1					
34906 ÷ 11111	= 3.14	15					
349066 ÷ 111111	= 3.14	15	9				
3490658 ÷ 1111111	= 3.14	15	92				
34906585 ÷ 11111111	= 3.14	15	92	6			
349065850 ÷ 111111111	= 3.14	15	92	65			
3490658504 ÷ 1111111111	= 3.14	15	92	65	3		
34906585039 ÷ 11111111111	= 3.14	15	92	65	35		
349065850398 ÷ 111111111111	= 3.14	15	92	65	35	8	
3490658503988 ÷ 1111111111111	= 3.14	15	92	65	35	89	
34906585039886 ÷ 11111111111111	= 3.14	15	92	65	35	89	7

Pythagoras: Again, tell me how you got each of those digits in the set of ratios?

JJA: It took months of experimenting to find my Mathematical Rosetta Stone. Once I found it, I was able to use my computer to discover a unique ratio that fit <u>every</u> <u>single</u> <u>variation</u> in **pi**.

Pythagoras: So, no matter how many integers after the decimal, even up to 51 billion—wherever we stop there is a unique ratio that will reproduce all those digits?

JJA: That is absolutely correct. For example, if we stop at **pi** = 3.1415926, then we can generate all those numbers with a unique ratio of 34906585 divided by 11111111.

Pythagoras: Astounding! What's the next constant you will try?

JJA: How about **e**?

Pythagoras: You mean to tell me this works for **e** also? The constant known as "**e**" is important because it appears in scores of scientific equations.

JJA: I was also astounded when the Mathematical Rosetta Stone deciphered **e** into a perfect set of ratios.

Pythagoras: Show me!!

JJA: Ok, here it is:

Set of Ratios	
	e = 2.71 82 81 82 84 59 04 5...
30 ÷ 11	= 2.7
301 ÷ 111	= 2.71
3020 ÷ 1111	= 2.71 8
30202 ÷ 11111	= 2.71 82
302031 ÷ 111111	= 2.71 82 8
3020312 ÷ 1111111	= 2.71 82 81
30203131 ÷ 11111111	= 2.71 82 81 8
302031314 ÷ 111111111	= 2.71 82 81 82
3020313142 ÷ 1111111111	= 2.71 82 81 82 8
30203131427 ÷ 11111111111	= 2.71 82 81 82 84
302031314272 ÷ 111111111111	= 2.71 82 81 82 84 5
3020313142732 ÷ 1111111111111	= 2.71 82 81 82 84 59
30203131427322 ÷ 11111111111111	= 2.71 82 81 82 84 59 0
302031314273227 ÷ 111111111111111	= 2.71 82 81 82 84 59 04
3020313142732272 ÷ 1111111111111111	= 2.71 82 81 82 84 59 04 5

Pythagoras: Again, I am impressed. I should say I am thrilled! If we had known this, my school of mathematics would have continued for centuries instead of only lasting 500 years!

JJA: You always advised your students that if something in mathematics is successful, that they should continue to work with it. So, that is exactly what I'm going to do next.

Summary

The famous "irrational" number for **pi**, which equals 3.14159265358979323846... has entertained math buffs for several thousand years. The largest number of digits after the decimal that have been calculated so far is more than 50 billion—with no recurring pattern in all those unending numbers. Now, for the first time, my Mathematical Rosetta Stone decodes **pi** to find a ratio of whole numbers for every digit before and after the decimal. The Mathematical Rosetta Stone also decodes another famous constant, the "irrational" number for **e**.

CHAPTER 6
The Mathematical Rosetta Stone decodes Euler's Constant and the Golden Mean

Pythagoras: Euler's Constant is a <u>different</u> kind of "irrational" compared with $\sqrt{2}$, pi or e.

JJA: How so? Euler's Constant is .577215664901532...

Pythagoras: It does <u>not</u> start with a whole number like the others. It starts with a decimal point followed by an unending "patternless" series of integers. This may be a more difficult test of the Mathematical Rosetta Stone. What do you think?

JJA: I don't think so. But let's find out. Euler's Constant is .577215664901532...

Set of Ratios	Euler's Constant = .57 72 15 66 49 01 53 2...
11 ÷ 19	= .57
111 ÷ 192	= .57
1111 ÷ 1923	= .57 7
11111 ÷ 19247	= .57 72
111111 ÷ 192494	= .57 72 1
1111111 ÷ 1924949	= .57 72 15
11111111 ÷ 19249496	= .57 72 15 6
111111111 ÷ 192494967	= .57 72 15 66
1111111111 ÷ 1924949683	= .57 72 15 66 4
11111111111 ÷ 19249496828	= .57 72 15 66 49
111111111111 ÷ 192494968286	= .57 72 15 66 49 0
1111111111111 ÷ 1924949682888	= .57 72 15 66 49 01
11111111111111 ÷ 19249496828895	= .57 72 15 66 49 01 5
111111111111111 ÷ 192494968288957	= .57 72 15 66 49 01 53
1111111111111111 ÷ 1924949682889593	= .57 72 15 66 49 01 53 2

Pythagoras: I'm dazzled! Let's take another look at the nature of decimals.

JJA: We established that decimals are not numbers in the sense of the natural numbers such as 1, 2, 3, 4 and so forth.

Pythagoras: Natural numbers to me represent a collection or in your modern parlance a "set."

JJA: So if we are talking about stones, the natural numbers would be the collection or the set containing 1 stone or 2 stones or 3 stones...

Pythagoras: Right

JJA: A decimal, however, is not a natural number. Each change in the decimal represents a <u>change in scaling</u>. With Euler's Constant for instance, .5 would be a scale of ten, .57 changes the scale to one hundred and .577 changes the scale again to one thousand.

Pythagoras: What the Mathematical Rosetta Stone shows is that each expansion of the decimal has a corresponding ratio. There is a one-to-one correspondence

JJA: Exactly.

Pythagoras: If I had know this, my school of mathematics would still be thriving today in the 21st century.

JJA: Should we try another application of the Mathematical Rosetta Stone?

Pythagoras: Absolutely! Let's try the Golden Mean which is equal to 1.61803398874989418...

JJA: As you know, the Golden Mean (also known as the Golden Rectangle) has a rich history. The ancient Egyptians referred to it as a "sacred ratio" in the Rhind papyrus, (circa 1650 B.C.).

Pythagoras: I know that this "sacred ratio" was used to construct the Great Pyramid at Giza.

JJA: Yes. The ratio of the height to half the width of the base is almost exactly 1.618. And the ancient Greeks used the Golden Rectangle to construct their splendid architecture - such as the Parthenon.

Pythagoras: I'll let you in on a little secret... even I used that marvelous ratio to construct the pentagram, one of the symbols of our brotherhood.

JJA: Let's take a look at the results:

<u>Set of Ratios</u>   The Golden Mean = 1.61 80 33 98 87 49 89 4...

18 ÷ 11	= 1.6
179 ÷ 111	= 1.61
1798 ÷ 1111	= 1.61 8
17978 ÷ 11111	= 1.61 80
179781 ÷ 111111	= 1.61 80 3
1797815 ÷ 1111111	= 1.61 80 33
17978155 ÷ 11111111	= 1.61 80 33 9
179781554 ÷ 111111111	= 1.61 80 33 98
1797815543 ÷ 1111111111	= 1.61 80 33 98 8
17978155430 ÷ 11111111111	= 1.61 80 33 98 87
179781554305 ÷ 111111111111	= 1.61 80 33 98 87 4
1797815543055 ÷ 1111111111111	= 1.61 80 33 98 87 49
17988155430554 ÷ 11111111111111	= 1.61 80 33 98 87 49 8
179881554305544 ÷ 111111111111111	= 1.61 80 33 98 87 49 89
1798815543055438 ÷ 1111111111111111	= 1.61 80 33 98 87 49 89 4

Pythagoras: I'm ecstatic! Tell me, will the Mathematical Rosetta Stone work with other physical constants in science?

JJA: The 28th edition of CRC Standard Mathematical Tables lists 23 physical constants. Here are three examples:

The equatorial radius of the earth = 6378.388 km = 3963.34 miles (statute).

The density of mercury at 0° C = 13.5955 g/ml.

The wavelength of the orange-red spectral line for Krypton 86 = 6057.802 Å.

Pythagoras: All of the numbers in the physical constants are thought to be "irrational" numbers

JJA: Yes, but, the patient reader will discover that each of the 23 physical constants listed will yield a set of whole number ratios when decoded with the Mathematical Rosetta Stone.

Pythagoras: This is sensational. Let's try something really crazy. Let's find out whether— and I must sound insane—let's explore whether the Mathematical Rosetta Stone can locate a pattern of ratios with any series of random numbers!

JJA: Why not?

Pythagoras: How will we test the hypothesis?

JJA: From the Table of Random Units, let's start by selecting a series of 16 randomly arranged digits.

Pythagoras: All Right. I'm looking at the table now, and I am going to select this set of random numbers:

1 2 2 4 3 7 9 9 8 8 2 6 0 1 0 5

JJA: Now, using the Mathematical Rosetta Stone, let's give the decoding a stringent test.

Pythagoras: I like that. Try to decode for three different patterns.
First try: .1224379988260105
next try: 1.224379988260105
and finally try: 12.24379988260105

JJA: Here are the results for .1224379988260105...

Set of Ratios   random number = .12 24 37 99 88 26 01 05

11 ÷ 91   = .12
111 ÷ 909   = .12 2
1111 ÷ 9075   = .12 24
11111 ÷ 90750   = .12 24 3
111111 ÷ 907495   = .12 24 37
1111111 ÷ 9074890   = .12 24 37 9
11111111 ÷ 90748884   = .12 24 37 99
111111111 ÷ 907488790   = .12 24 37 99 8
1111111111 ÷ 9074887877   = .12 24 37 99 88
11111111111 ÷ 90748878760   = .12 24 37 99 88 2
111111111111 ÷ 907488787600   = .12 24 37 99 88 26
1111111111111 ÷ 9074887876019   = .12 24 37 99 88 26 0
11111111111111 ÷ 90748878760260   = .12 24 37 99 88 26 01
111111111111111 ÷ 907488787602649   = .12 24 37 99 88 26 01 0
1111111111111111 ÷ 9074887876026510 = .12 24 37 99 88 26 01 05

Pythagoras: Fantastic. It worked! Now, try this pattern when the decimal is moved one place to the right, like this: 1.224379988260105...

JJA: Here are the results for 1.224379988260105

Set of Ratios		random number = 1.22 43 79 98 82 60 10 5
14 ÷ 11		= 1.2
136 ÷ 111		= 1.22
1360 ÷ 1111		= 1.22 4
13604 ÷ 11111		= 1.22 43
136041 ÷ 111111		= 1.22 43 7
1360421 ÷ 1111111		= 1.22 43 79
13604221 ÷ 11111111		= 1.22 43 79 9
136042220 ÷ 111111111		= 1.22 43 79 98
1360422209 ÷ 1111111111		= 1.22 43 79 98 8
13604222091 ÷ 11111111111		= 1.22 43 79 98 82
136042220918 ÷ 111111111111		= 1.22 43 79 98 82 6
1360422209178 ÷ 1111111111111		= 1.22 43 79 98 82 60
13604222091779 ÷ 11111111111111		= 1.22 43 79 98 92 60 1
136042220917789 ÷ 111111111111111		= 1.22 43 79 98 82 60 10
1360422209177894 ÷ 1111111111111111		= 1.22 43 79 98 82 60 10 5

Pythagoras: This blows me right out of my toga. This is a purely random number that is related to nothing at all and yet, there is an underlying pattern! Even "randomness" has a pattern. This tells me that there is no such thing as randomness. It is an illusion! Isn't this what Dr. Einstein believed?

JJA: I'm going to interview Dr. Einstein later. I'll ask him. Right now, get ready for the final test—in which we will move the decimal two places to the right, as you requested.

Set of Ratios   random number = 12.24 37 99 88 26 01 10

135 ÷ 11	= 12.2
1359 ÷ 111	= 12.24
13603 ÷ 1111	= 12.24 3
136040 ÷ 11111	= 12.24 37
1360420 ÷ 111111	= 12.24 37 9
13604220 ÷ 1111111	= 12.24 37 99
136042219 ÷ 11111111	= 12.24 37 99 8
1360422208 ÷ 111111111	= 12.24 37 99 88
13604222090 ÷ 1111111111	= 12.24 37 99 88 2
136042220917 ÷ 11111111111	= 12.24 37 99 88 26
1360422209177 ÷ 111111111111	= 12.24 37 99 88 26 0
13604222091778 ÷ 1111111111111	= 12.24 37 99 88 26 01
136042220917789 ÷ 11111111111111	= 12.24 37 99 88 26 01 1
1360422209177893 ÷ 111111111111111	= 12.24 37 99 88 26 01 10

Pythagoras: As Archimedes would say, "This is delicious!" So what is the grand conclusion from this spectacular demonstration?

JJA: The conclusion is this: Even random numbers (of any length) are associated with a set of whole number ratios in which 11 and its expansion are located in either the numerator or the denominator.

Pythagoras: Well, you showed earlier that we are not limited to 11. The Mathematical Rosetta Stone will decode with 22, 33, and so forth

JJA: Thank you. Yes, you are correct.

Pythagoras: There are then, no random numbers. All natural numbers, in any arrangement, have a pattern.

JJA: That is what I believe is true. The historical concept of patternless numbers is a myth.

Summary

Pythagoras wonders whether Asher's Mathematical Rosetta Stone will work for Euler's Constant which does not start with a whole number and looks like this: .577215664901532... Not only does it work for Euler's Constant, but also the famous "irrational" number for the Golden Mean, which equals 1.61803398874989418...

Pythagoras wants to know whether the decoding will work for the scores of physical constants (all thought to be "irrational" numbers) listed in CRC Standard Mathematical Tables. The answer is, yes.

Pythagoras then wants to explore "random" numbers. Can the Mathematical Rosetta Stone demonstrate a pattern of ratios even for so-called "random" numbers? The answer will surprise you. It surprised Pythagoras.

CHAPTER 7

Dr. Albert Einstein believes that "randomness" is an illusion. Perhaps we can prove him right!

A conversation with Dr. Einstein

Dr. Einstein: You presented a concept that attracted my attention.

JJA: What was that?

Dr. Einstein: The concept that "randomness" is an illusion since any so-called random sequence of numbers can be converted into Pythagorean ratios. The existence of irrational numbers turns out to be an illusion. There is no such thing as "irrational" numbers. All numbers are rational. I love it! Mathematics like science obeys Newtonian clockwork.

JJA: Can you suggest any other interesting applications of the Mathematical Rosetta Stone?

Dr. Einstein: I have a "weird and wonderful" idea. Can we perform arithmetic with "random" sequences of numbers to create a new set of "random" numbers? Then perform the same arithmetic on the Pythogorean ratio decoded from the original sequence of "random" numbers. Will we get the identical results?

JJA: Fantastic! Let's try it.

Dr. Einstein: I suggest that we start with the square root of 2, which has double handicap of being both "random" and "irrational."

JJA: Applying the Mathematical Rosetta Stone, we get:

Pythagorean Ratios	$\sqrt{2}$ = **1.4142**
16/11	= 1.4
157/111	= 1.41
1571/1111	= 1.414
1513/11111	= 1.4142

Dr. Einstein: Add an odd number to $\sqrt{2}$. Let's say 3.

JJA: Right. $\sqrt{2}$ = 1.4142 plus 3 will equal **4.4142** Now, let's see what happens when we add 3 to the Pythagorean ratios for the square root of 2.

Pythagorean Ratios $\sqrt{2}$ = 1.4142 plus 3 = **4.4142**

16/11 + 3/1 =
16/11 + 33/11 = 49/11   = 4.

157/111 + 3/1 =
157/111 + 333/111 = 490/111   = 4.4

1571/1111 + 3/1 =
1571/1111 + 3333/1111 = 4904/1111   = 4.41

15713/11111 + 3/1 =
15713/11111 + 33333/11111 = 49046/11111   = 4.4142

Dr. Einstein: It's working.

JJA: Now what?

Dr. Einstein: Let's try multiplying $\sqrt{2}$ by 3

JJA: Here we go: √2 times 3 = **4.242**...

<u>Pythagorean Ratios</u>　　　√2 = 1.4142 times 3 = **4.242**
16/11 times 3/1 = 48/11　　　　　　　　　　= 4.
157/111 times 3/1 = 471/111　　　　　　　　= 4.2
1571/1111 times 3/1 = 4713/1111　　　　　　= 4.24
15713/11111 times 3/1 = 47139/11111　　　　= 4.242

Dr. Einstein: Voilá! Now try dividing √2 = 1.4142 by 3.

JJA: √2 = 1.4142, and if we divide by 3 we get **0.4714**

<u>Pythagorean Ratios</u>　　　√2 = 1.4142 divided by 3 = **0.4714**
16/11 divided by 3/1 =
16/11 times 1/3 = 16/33　　　　　　　　　　= .4
157/111 divided by 3/1 =
157/111 times 1/3 = 157/333　　　　　　　　= .47
1571/1111 divided by 3/1 =
1571/1111 times 1/3 = 1571/3333　　　　　　= .471
15713/11111 divided by 3/1 =
15713/11111 times 1/3 = 15713/33333　　　　= .4714

Dr. Einstein: We hit the target again. I'm convinced. Using Pythagorean ratios decoded with your Mathematical Rosetta Stone, we can successfully apply arithmetic with so-called "random" numbers which tells me that the concept of randomness is an illusion.

"Randomness" and "probability" are wastebasket terms for the unknown. When we demonstrate true cause-effect, there is no randomness. There is no probability; both terms disappear.

JJA: Any other examples?

Dr. Einstein: The field of medicine is rich with examples. For instance, Dr. Walter Reed was called in to solve the problem of malaria that was killing one of every two workers on the Panama Canal project.

JJA: As I recall, the French abandoned the project because the risk of death was so high that no one wanted to work in Panama.

Dr. Einstein: President "Teddy" Roosevelt with his marvelous American self-confidence said that "We can do it! We will do it!" But first, somehow, the problem of malaria had to solved.

JJA: The prevailing medical opinion was that the source of the malaria was garbage everywhere on the streets of Panama.

Dr. Einstein: Dr. Reed argued that the mosquito was the source of malaria rather than garbage.

JJA: He was over-ruled with an expensive project that removed the garbage from Panama's streets.

Dr. Einstein: It didn't work. Still one of every two workers was a victim of malaria. The mystery was, we didn't know who would contract the illness. So we were into probabilities. The probability was 1 chance in 2 that a worker would be affected. We could not predict with precision for an individual worker. Here we have chance, probability, and uncertainty.

These to me are labels that I call wastebasket terms for any unknown cause and effect.

JJA: When the USA finally decided, with great reluctance, to test Dr. Reed's hypothesis...

Dr. Einstein: They sprayed all of Panama to kill mosquitos and the incidence of malaria fell almost immediately to near zero. An exciting illustration of what happens when we discover the true cause and effect of a phenomenon.

JJA: One more illustration.

Dr. Einstein: Blood transfusions have an interesting history. The first attempt was to transfer blood from animals to humans. Some people survived and regained their health, but others did not.

JJA: What happened next?

Dr. Einstein: Next, medical practitioners tried transferring blood from a healthy person to someone who was ill.

JJA: And the results?

Dr. Einstein: Some patients survived and recovered from their illness, but many people died. This problem lingered in the gray area of probability.

JJA: How so?

Dr. Einstein: Each procedure had a certain probability of success. Animal to human blood transfer had a lower probability of success than a human to human transfer.

JJA: Then came a breakthrough.

Dr. Einstein: Yes. Blood typing.

JJA: Blood typing has the exciting benefit of almost perfect prediction.

Dr. Einstein: Right. The ***probability*** of success was replaced with ***certainty*** of success. Again and again, when we find the true cause and effect, probability disappears.

Summary

Dr. Einstein is excited about my Mathematical Rosetta Stone because the results suggest that "randomness" is an illusion. He always believed that science is not possible if nature is the end product of a chance coalition of events. As Dr. Einstein was fond of saying, "God is certain. If God was uncertain, then scientific discoveries would not be possible."

Dr. Einstein asked to play with "random" sequences of numbers using arithmetic. The results are again surprising.

CHAPTER 8

In 1913 banner headlines across the world screamed in huge black print: *"Einstein overturns Newton's laws of gravity."*

Dr. Albert Einstein sets the record straight

Dr. Einstein: Newton left us with a puzzle that has baffled physicists for 300 years. He gave us a precise mathematical relationship that allowed us to predict the attraction of planets millions of miles away from each other. Without that contribution, space travel would still be a Jules Verne fantasy.

JJA: The popular view is that you overturned Newton's mathematics about gravity.

Dr. Einstein: Nonsense. That is a newspaper headline in 1913 or thereabouts. A headline that captured the public's imagination. I simply expanded upon Newton's sensational work.

JJA: What was the puzzle?

Dr. Einstein: Even though Newton's work allows us to predict the movement of planets and the attraction of celestial objects, the mystery remained: What was the mechanism that explained this amazing action at a distance?

JJA: Let's start with your concept of space. How is your three-dimensional fabric of space different from the now defunct concept of "ether" - the invisible substance that was once thought to pervade all of space?

Dr. Einstein: There is a similarity. Ether was conceived as a medium and so is my concept of space. The difference is that my concept of three-dimensional space is an explanation for the mechanism of gravity.

JJA: So in a vacuum jar, when the oxygen is removed, the jar is not empty.

Dr. Einstein: It is not empty. You cannot empty out the fabric of space. Space is an entity. It is something. It is not nothing. Each of us, for instance, as we move about, are making a slight indentation in the fabric of space.

JJA: How does your concept of "space" explain gravity?

Dr. Einstein: The analogy I like is this: Imagine a thin rubber sheet stretched tightly on a frame. Now if you drop a bowling ball in the middle, there will be an indentation in the rubber around the heavy bowling ball. Next, roll a marble in the direction of the bowling ball. The marble represents the earth and the bowling ball represents the sun.

JJA: The marble will roll straight until it gets to the indentation.

Dr. Einstein: Yes, then the marble will go over the edge and begin to spiral around the bowling ball. That is the earth orbiting around the sun.

JJA: How about the moon?

Dr. Einstein: The moon is a celestial mass that rolled into the indentation made by the earth and began to spiral or orbit the earth.

JJA: And "black holes"?

Dr. Einstein: A black hole is created by an object perhaps a million times the mass of our sun. It is like the Empire State Building dropped in the middle of our thin rubber sheet. There is an indentation the size of a small grand canyon. Any object coming close to this huge indentation begins to spiral at millions of miles per hour. If the earth were to come close enough, it would be sucked swirling into the gravity well and reduced to the size of a marble.

JJA: And light?

Dr. Einstein: Light coming into the black hole will be bent and curve around, and unable to exit.

JJA: Fascinating. Now let's explore time. You feel that "time" is also an entity. It is something. If space has three dimensions, then time is the fourth dimension.

Dr. Einstein: Yes.

JJA: With great respect for you and your achievements, I would like to challenge the notion that "time" is something.

Dr. Einstein: Continue. This is interesting. What is "time" to you?

JJA: An abstraction.

Dr. Einstein: Representing what?

JJA: Representing a cluster of measurements for events in a network that are happening either sequentially or simultaneously. There is a net of events moving horizontally or vertically.

Dr. Einstein: Hummm. Let's start with something simple. Let's start with inches, meters, and fathoms. What are they?

JJA: Measurements for distance—that is, measurements for the distance between two points.

Dr. Einstein: Agreed. How about pounds and kilograms?

JJA: Measurements for weight or mass.

Dr. Einstein: All right. Now for "time." Let's examine historical representations for time. What is the earliest measurement for time?

JJA: Perhaps sun up and sun down. Others would be counting the appearance of the moon, the change of seasons, gestation, the change in physical appearance... Remember Galileo? He took his pulse to measure the "time" a hanging candle in church moved in an arc back and forth.

Dr. Einstein: Then into mechanisms such as the sun dial, the water clock and the pendulum, which was developed from Galileo's observation of the swinging candle.

JJA: As we go through history, the measurements go from crude to sophisticated.

Dr. Einstein: Crude to sophisticated meaning we started with large intervals and step by step invented ways to represent smaller and smaller intervals of "time." The smallest interval can perhaps be measured with an atomic clock.

JJA. I don't have an atomic clock but I do have strapped to my wrist a line moving around the circumference of a circle which measures rather small intervals.

Dr. Einstein: Yes, I have one of those strapped to my wrist also.

JJA: Now to the nature of time. To me "time" is not something but rather an abstraction representing events such as revolutions of the earth and other planets. They are happening either in sequence or simultaneously. Each revolution of the earth is in a sequence which we call years, but simultaneously, other planets are also revolving. "Time" does not exist. There is no such thing as "time."

Dr. Einstein: How is the clock different from other measurements such as inches, meters, or fathoms?

JJA: Inches, meters, fathoms and pounds are one dimensional while the clock represents two dimensions: sequence and simultaneity of events. The clock allows us to predict events that occur in sequence or happen simultaneously.

Dr. Einstein: Give me some homey examples.

JJA: In one hour, you will be at the lake in Princeton preparing your sailboat for launching. At this moment while we are having this conversation, your wife is home preparing your lunch.

Dr. Einstein: I see your point.

JJA: Another way of looking at the non-existing entity of "time." If we only strapped a circle on our wrists because we are fascinated with a line revolving around the circumference of the circle, we might be considered eccentric if not schizophrenic.

Dr. Einstein: The line moving around the edge of a circle allows prediction.

JJA: That is the value. If it did not allow us to forecast events that are important to us, we would not strap it to our wrists.

Dr. Einstein: And for convenience and reliability, we prefer the line moving around a circle on our wrists than a burning candle or water clock or sand clock.

JJA: Yes. I like the concept. It appeals to the maverick in me. There is no such thing as "time." Time is a reification. "Time" is so important to us that we have transformed it into a creature, a thing, an entity.

Summary

Dr. Einstein clarifies his contribution to Newton's laws of gravity. Sir Isaac Newton discovered equations that predict with precision the attraction of one planet to another separated by huge distances. How "action at a distance" works has been a mystery until Dr. Einstein provided a model that explains the mechanism. Then I explore the nature of "time" with Dr. Einstein. He feels that "time" is an entity —something real while I present an alternative explanation.

CHAPTER 9
René Descartes reveals
how he made his startling discovery of a mysterious
"Atlantis" connecting Algebra and Geometry—
a connection that mathematicians for centuries
believed did not exist.

JJA: Certainly if there had been a Nobel Prize in your time, you would have been honored with the prize no matter who in history was competing, including Dr. Einstein.

Descartes: You know that Alfred Nobel, for reasons known only to himself, excluded the field of mathematics from Nobel Prize consideration.

JJA: Yes, I know. But the implications of your discovery imprinted all other fields of science. It is as if science was moving in one direction and your contribution suddenly made everything turn ninety degrees.

Descartes: I appreciate that but I cannot take credit. This was a revelation rather than a discovery. It was a true epiphany.

JJA: Tell me how it came about.

Descartes: I knew that for hundreds of years the prevailing belief was that there is no connection between algebra and geometry.

JJA: You did not accept that powerful belief.

Descartes: No.

JJA: Why not?

Descartes: Because I am convinced that the Creator likes simple patterns. Algebra and geometry had to be one unified system. It just had to be. All of mathematics has to be one unified system.

JJA: The fragments and complexities are an illusion?

Descartes: Absolutely.

JJA: Then what?

Descartes: I began a search to find the "answer."

JJA: The story is that the "answer" came to you in a dream.

Descartes: It did. One night while I was in a deep sleep, the Angel of Truth came to me and whispered the secret connection between algebra and geometry. It was so simple that I awoke with a jolt and with a trembling hand, I scribbled it down on a scrap of paper.

JJA: If we had that scrap of paper today, it would be a priceless treasure. What exactly was on that paper?

Descartes: The secret of a straight line.

JJA: Please explain.

Descartes: Well, let's look at a line like this:

Tell me, how would you measure it?

JJA: With a ruler.

Descartes: Could you determine the distance between 4 and 10 on the line?

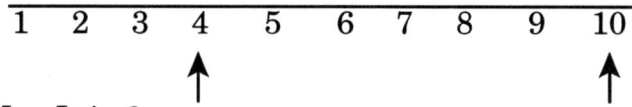

JJA: Yes. It is 6.

Descartes: How do you know?

JJA: Because I subtracted 4 from 10. Nothing to it.

Descartes: You subtracted the beginning point from the end point.

JJA: Yes.

Descartes: The mystery: How can we represent that length on a line for <u>all</u> numbers, not just specific numbers? We needed a way to measure the length of a line for <u>all</u> numbers.

JJA: You wanted to move up in abstraction from specific numbers to letters that would represent all numbers.

Descartes: Exactly.

JJA: So, on that scrap of paper, you wrote?

Descartes: I wrote X_1 for one point on the line and X_2 for another point on the line like this..

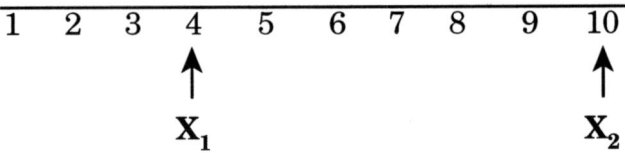

JJA: Then the length of the horizontal line is X_2 minus X_1 or expressed differently, $X_2 - X_1$

Descartes: And for a vertical line, can you tell me the distance between, let us say, 3 and 7 ?

```
 | 10
 |
 |  9
 |
 |  8
 |
 |  7  ←
 |
 |  6
 |
 |  5
 |
 |  4
 |
 |  3  ←
 |
 |  2
 |
 |  1
```

JJA: The answer is 4. I subtracted 3 from 7 to get 4.

Descartes: Suppose the 7 was Y_2 and the 3 was Y_1 as in the next line, can you still find the distance?

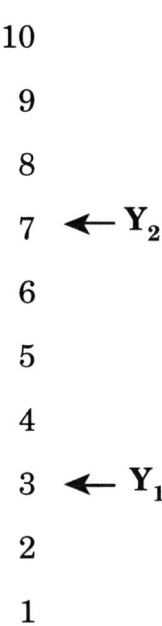

JJA: Yes. I will subtract the beginning point from the end point to get $Y_2 - Y_1$

Descartes: This means that we can express the distance across and up on geometric figures such as triangles.

JJA: So you have translated geometry into algebra. But what does it all mean?

Descartes: Once horizontal and vertical lines can be expressed in letters, we can explore geometric relationships using algebra, and you want to know the significance of this?

JJA: Yes, definitely. I presume that if you can go from geometry to algebra, the reverse is also true...

Descartes: If we measure horizontal and vertical lines with numbers only, we are limited to geometry. But once we move up in abstraction from numbers to letters that represent all numbers, we can play with algebra. Letters representing points on horizontal and vertical lines allow us to move back and forth from geometry to algebra and algebra to geometry.

JJA: I would appreciate an example, but before we explore this, I'm curious. Sir Isaac Newton complained that you concealed this "simple" idea in obscure and convoluted language. He became utterly exhausted after reading just one paragraph! He thought that you were mean-spirited to cloak your thinking with such obscure writing.

Descartes: I did not intend to be difficult. Remember that it was during the time of the inquisition. The search for heretics was everywhere—led by a few cruel men of the cloth who would not hesitate to disfigure by burning, twisting, bending and cutting the flesh to save a person's soul. I saw what Galileo in Italy endured and I did not want the same fate!

JJA: The ambiguity in your writing was to escape the scrutiny of the censors?

Descartes: Exactly.

JJA: Why would mathematical concepts be perceived as heresy?

Descartes: Who knows? I didn't know how creative the inquisitors might be in the interpretation of a new idea. I didn't know what they might see that would threaten their belief in Genesis, for example.

JJA: How could concepts in geometry or algebra be threatening to biblical messages?

Descartes: Well, Galileo looked through his telescope and saw other planets. That scared the clergy. It threatened their belief that the earth is the center of the universe.

JJA: I understand now why you would be anxious. If Newton had difficulty understanding your writing, then you were certainly safe from everyone else!

Descartes: So where do we go from here?

JJA: Back to our discussion of algebra. What does algebra have that numbers do not have? Why convert from numbers to letters?

Descartes: Good question. That question goes to the heart of algebra. The answer to that question will tell us why algebra is so valuable?

JJA: Every school child wants to know.

Descartes: Here is my view—when we measure lines with numbers, for example, <u>input</u> <u>will</u> <u>equal</u> <u>output</u>. Information is limited.

JJA: And if we convert numbers into letters to measure lines, then what?

Descartes: Then output will be <u>more</u> than input. It is a kind of mathematical alchemy. We experiment with lead and iron to get gold.

JJA: I don't understand. How does it work?

Descartes: Algebra is a Rosetta Stone that lets us decode hidden messages from the Creator about how the universe works. Algebra may even tell us how the universe was created.

JJA: I can see that statements like that could get you convicted of heresy.

Descartes: Ahhh... now you are beginning to understand.

JJA: What can algebra do that arithmetic cannot? What is the magic of algebra?

Descartes: Let's explore an example from the amazing Euclid, whose book, *Elements of Geometry*, has, as you know, been reprinted in more than 2,000 editions. Using only arithmetic, how can anyone demonstrate that there is no ratio, no fraction, that represents the square root of 2?

JJA: To find a ratio for the square root of 2, we could try trial-and-error, but that would take forever with no guarantee that the ratio is not hiding somewhere in the trillionth ratio.

Descartes: That's the limitation of using arithmetic to find a solution. We would have to examine every possible combination of integers 1, 2, 3, 4... into infinity—and <u>infinity</u> <u>does</u> <u>not</u> <u>exist</u>.

JJA: Infinity doesn't exist?

Descartes: No. It is not a destination. It is a process. There is no ending point. Hence, it does not exist. For example, start with numbers such as 1, 2, 3, 4... What is the last number? It does not exist. Therefore, infinity simply refers to a continuation of 5, 6, 7... and so on forever.

JJA: If arithmetic had a "last number," it would be possible to find that illusive ratio for the square root of 2.

Descartes: Correct. The fly in the mathematical ointment is that mysterious, non-existent, infinity. Only algebra is a code that allows us to decipher mathematical mysteries that involve "infinity."

JJA: Can you show us how?

Descartes: Sure. But first I want to comment about your conversation with Cauchy. I like your interpretation that negative numbers in multiplication have a *double meaning* with the first number signaling direction, either right or left on the number line, and the second number instructing us to continue in the same direction or reverse directions. It has a symmetry that appeals to me.

JJA: These two arithmetic operations of addition and multiplication seem to be parallel but this is not quite true. The common perception is that multiplication is simply repeated addition.

Descartes: Multiplication is not repeated addition?

JJA: Only in the special case of positive numbers is multiplication simply repeated addition. Negative numbers in "addition" such as -1, -2, -3, etc. do not have the same meaning as negative numbers in "multiplication."

Descartes: Can you give some examples?

JJA: Yes. For instance, in "addition," a -2 means to move two steps to the left of zero on the number line. If we add another -2, we move two more steps in the same direction and arrive at -4. When adding (-2) + (-2), both -2 and -2 have the "same" meaning.

If we multiply a -2 times a -2, the first minus sign means to move to the left, and the second minus sign means to reverse direction. Hence, in multiplication, -2 and -2 each have a "different" meaning.

Descartes: How does this all apply to fractions?

JJA: If multiplication is simply repeated addition, then we have another mystery. Why do we "increase" the magnitude of the end-product when we add fractions but "decrease" when we multiply fractions?

Descartes: Let's see—when as we add more and more fractions, the result is a larger and larger number. I need a "for instance." Let's try this:

1/2 + 1/2 = **1**
1/2 + 1/2 +1/2 = **1.5**
1/2 + 1/2 + 1/2 + 1/2 = **2**

I agree. When we add fractions, the result is an increase.

JJA: Now try the same series with multiplication.

Descartes: Here we go:

1/2 times 1/2 = **1/4**
1/2 time 1/2 times 1/2 = **1/8**
1/2 times 1/2 times 1/2 times 1/2 = **1/16**

When we multiply fractions, the result is a decrease. Good. Now what? How do you explain this apparent contradiction?

JJA: I believe that multiplication involves a two-step operation. Step one is "copy" and step two is "add."

Descartes: Please demonstrate.

JJA: For example: (1)(1) means to copy 1 once and add the result to zero. Thus, 1 + 0 = 1

(1)(2) means to copy 2 once and add the result to zero. Thus, 2 + 0 = 2

Or, it can be done in reverse.
(2)(1) means to copy 1 twice and add the result to zero. Thus, 1 + 1 + 0 = 2

Descartes: Try one more example.

JJA: (2)(2) means to copy 2 twice and add the result to zero. Thus, 2 + 2 + 0 = 2

Notice that two step process of copying and then adding, has been invisible in our historical perception of what multiplication is.

Descartes: But how does your *copy and add* interpretation of multiplication explain the asymmetry between addition and multiplication of fractions?

JJA: If multiplication is just repeated addition, then the results should increase for both addition and multiplication of fractions.

Descartes: I think so. But it doesn't. Hence a contradiction. What is the explanation?

JJA: As I mentioned, in my model of multiplication, I believe there are two steps: *copy and add*.

If we ask school children to multiply 1/2 times 1/2, they will write down 1/4 without hesitation. But, if you ask them: "Why does 1/2 times 1/2 equal 1/4?" You will not see any hands waving in the air.

Descartes: They don't know why.

JJA: No they don't, but my *copy and add* offers an explanation. On the next page, I'll demonstrate how it works.

It goes like this.

1/2 times 1/2 means to *copy* 1/2 of 1/2 which is 1/4.

Picture a unit of one...

1

which is cut into two halves like...

1/2	1/2

If each half of the whole is cut into 1/2, we have four parts like this...

1/4	1/4	1/4	1/4

Now *copy* 1/2 of the first half and you get 1/4.

Descartes: 1/2 times 1/2 means to *copy* one of four parts which is 1/4. Since there is nothing to add it to, we *add* it to zero to get 1/4.

Try something that seems a bit more complex such as 1/2 of 1/4.

JJA: Again, you can picture the relationship when a whole is cut into two halves like this...

1/2	1/2

If we cut each half into four parts...

1/4	1/4	1/4	1/4

Now cut each fourth in half, which results in the whole being cut into eight parts, like...

1/8	1/8	1/8	1/8	1/8	1/8	1/8	1/8

Descartes: So if we *copy* 1/2 of 1/4, we get 1/8.

JJA: Notice that multiplying fractions *decreases* the product.

Descartes: Conclusion: If we limit our calculations to the special case of positive numbers, "multiplication is indeed repeated addition." But for negative numbers and fractions, the rule we were taught in school is false.

JJA: And my *copy and add* interpretation of multiplication explains why adding fractions will <u>increase</u> the end-product, but multiplication will <u>decrease</u> it.

Descartes: Not bad. Your *copy and add* theory seems to hold up for the multiplication of fractions. You are slowly convincing me that the missing link in understanding multiplication is the "copy" operation.

JJA: Now let's get back to our discussion of infinity.

Descartes: Infinity does not exist. There is no such thing as infinity. It is not something anyone has ever experienced. It is not a destination. It is not a new world or new planet that is somewhere in the universe.

JJA: Infinity is like walking, sailing, or driving along the surface of the earth toward the horizon. We can never reach it because it is always moving beyond us.

You said that only algebra is a code that allows us to decipher mathematical mysteries involving infinity. Can you show us how? Can you explain this so that any young student can understand?

Descartes: I was thinking of Euclid's proof showing that if we look forever at numbers, we will never find a single ratio for the square root of 2 because it does not exist. But, this may be too subtle for a child to follow.

JJA: Let's see... You start with a premise—an assumption that a single ratio does exist to represent the square root of 2, then you use logic with the language of algebra to disprove the idea. Let's try and see what happens ...

Descartes: Euclid starts with the assumption that a single ratio represented by **a/b** does exit for the square root of 2.

JJA: Wait! By single ratio you mean, for instance, the $\sqrt{4}$ = 2/1 because 2/1 times 2/1= 4/1.

Descartes: Yes. All whole numbers are ratios such as 1/1, 2/1, 3/1, and so forth.

JJA: What's the next step?

Descartes: He also assumes that **a/b** have no prime factors in common.

JJA: Let me play the role of the student. What do you mean by "no prime factors in common?"

Descartes: Here are some examples of what I mean: 2/4 can both be divided by the prime of 2 which reduces the fraction to 1/2. 6/15 can both be divided by the prime of 3 to reduce the fraction to 2/3, and 15/25 can be divided by the prime of 5 to get a fraction of 3/5 and so forth.

JJA: In other words, **a/b** must be in the lowest possible terms.

Descartes: Correct. Now let's apply some algebra.
a/b = square root of 2 or

$$\boxed{a/b = \sqrt{2}}$$

Let's get **a** on one side only.

JJA: <u>Multiply the left side by **b**</u> because **b/b** = 1 and **a** times 1 = **a**.

Descartes: Good, but the equation is now out of balance.

JJA: Restore balance or symmetry by <u>multiplying the right side by **b**</u> to get:

$$\boxed{a = b\sqrt{2}}$$

Descartes: Next, we want to eliminate the square root, so <u>square both left and right sides of the equation</u>:

$$\boxed{a^2 = 2b^2}$$

JJA: Since 2 times any number is an even number, the right side must be an even number.

Descartes: And if the right side is even, then the left side is also even. Hence, **a**² is an even number.

JJA: Now let's look at **a**². If **a**² is even, then **a** must be even. Here are some examples:

a²	**a**
4	2
16	4
36	6

Descartes: If **a** is even then **a** = 2n for some integer. Hence, **a**² = (2n)² or 4n².

JJA: If we substitute 4n² for **a**², we get:

$$\boxed{4n^2 = 2\mathbf{b}^2}$$

Simplify to get:

$\boxed{2n^2 = \mathbf{b}^2}$ ("Simplify" means to divide both sides of the above equation by 2.)

Descartes: Hence, **b**² must be even and if so, then **b** must also be even.

JJA: Euclid started with the assumptions: (a) there is a single ratio, called **a/b** for the √2, and (b) the single ratio **a/b** is in the lowest possible terms. Hence, **a/b** cannot be an *even* integer over an *even* integer. Since we found **a/b** to be *even over even*, we have a contradiction.

Descartes: Just to be sure our conclusion is clear... Euclid showed that $\sqrt{2}$ = **a/b** = **even/even** as, for example: **2/4** or **4/6** which are not in the lowest possible terms.

JJA: Hence, there does not exist a single ratio in the *lowest possible terms* for $\sqrt{2}$.

Summary

For hundreds of years, mathematicians believed that there was no connection between algebra and geometry. René Descartes reveals how he discovered the mysterious connection that is the foundation of Analytic Geometry—a branch of mathematics that is the basis of engineering and technology. Descartes shows us the power of algebra in coping with the mysteries of infinity.

CHAPTER 10
Einstein's Concept of Space and Zeno's Paradox

A conversation with Dr. Albert Einstein

JJA: Dr. Einstein, you picture space as an entity.

Dr. Einstein: Yes, to me space is not empty. It is three dimensional. As we move around, we make imperceptible indentations in the fabric of space.

JJA: I am wondering about your model of space and Zeno's paradox. Perhaps your model gives us a solution that has puzzled philosophers and mathematicians for centuries.

Dr. Einstein: Tell me what you have in mind.

JJA: A few thousand years ago in ancient Greece, Zeno made this observation: When an archer shoots an arrow towards a target, where is the arrow at every instance? Since the arrow must be a "rest" at every instance, it cannot be nowhere.

Dr. Einstein: I am trying to get a clear picture of the arrow in flight. So, at instance A, I see the arrow and then at instance B, there is the arrow.

JJA: The puzzle is: where is the arrow between A and B? Common sense tells us that the arrow is in "motion."

Dr. Einstein: This cannot be. Otherwise, it would be nowhere. It cannot be nowhere. How did the arrow get from A to B?

JJA: Let's play with this possibility: I visualize three-dimensional space as a kind of honeycomb. Let's label each cell in the honeycomb with a number. Let's say that the arrow before it is propelled into motion by the archer is at rest occupying cells 1, 2, and 3.

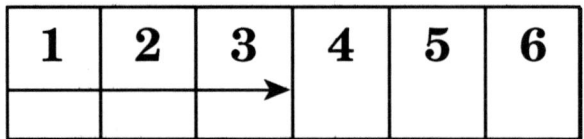

Dr. Einstein: The trick is to explain how the arrow then occupies cells 4, 5, and 6.

JJA: It may be that the arrow never occupies cells 4, 5, and 6.

Dr. Einstein: How can that be?

JJA: The force of the bow sets the arrow in motion. The arrow enters cell 4, but not the entire arrow, only one-third.

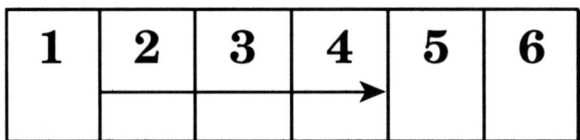

As the first third of the arrow leaves cell 4 and enters cell 5, the second third of the arrow is entering cell 4.

As the first part of the arrow enters cell 6, the last part of the arrow is entering cell 4.

1	2	3	4	5	6
					→

Dr. Einstein: Hence, the ***entire arrow*** is never in a different location on its flight.

JJA: Exactly.

Dr. Einstein: So, at the start before being set in motion, the arrow occupies multiple cells of 1, 2 and 3.

JJA: Yes.

Dr. Einstein: Then it moves, but only part of the arrow will occupy the adjacent cell 4, then move to cell 5 as another part of the arrow now occupies cell 4.

JJA: Yes. Once in motion, the ***entire arrow will never occupy adjacent cells*** until reaching the target where the entire arrow at rest will occupy let us say, cells 98, 99, and 100.

Dr. Einstein: Again let us consider this scenario. At the start, before being set in motion, the arrow occupies cells 1, 2 and 3. Then it moves and occupies cell 2, 3, and 4, then 3, 4, and 5. There is no unique location that is non-overlapping until the target is reached.

JJA: Once in motion, there is no discrete, non-overlapping location for the "entire" arrow.

Dr. Einstein: When the arrow is in motion, it will always overlap with a previous address. The overlap ceases when it again comes to rest at the target.

JJA: But, so what? Does this answer Zeno's paradox?

Dr. Einstein: While in flight, the entire arrow is never in a non-overlapping location until it is again at rest at the target.

JJA: There are only two places in space where the arrow is in non-overlapping locations: at the start before being set in motion, and at the end when it reaches the target.

Dr. Einstein: Hmmm. Interesting. I would like to play with that thought. It has implications. Zeno's paradox is an illusion but what is the illusion?

JJA: The illusion is that Zeno presented a "fact" to work with, but the fact turns out to be a fiction. He wants us to place the arrow in flight in non-overlapping locations at each instance, which is impossible. Then he asks: "How did the arrow jump from cells 1, 2, and 3 to cells 4, 5, and 6"?

Dr. Einstein: The answer is that the entire arrow never was in cells 4, 5, and 6 at any time in flight. Bravo!

Summary

It is fun to explore one of Zeno's famous paradoxes with Dr. Einstein. When an arrow is released from an archer's bow, where is it located at each moment in time? It cannot be in transition, which would make its location "nowhere." At every instance, the arrow has to be at rest somewhere. To explain the paradox, I use Dr. Einstein's concept that space is a real entity.

CHAPTER 11
What did Galileo, Newton, Maxwell, Faraday, and Einstein have in common?
A conversation with Dr. Albert Einstein

JJA: Did you replace Newton's laws of gravity?

Dr. Einstein: Not at all. I just expanded his concept of gravity to include more observations. His famous inverse square law still holds, up to a point, then the indentation in space produces some bending which my model can explain.

JJA: The model that can explain more data is a simpler model of the universe.

Dr. Einstein: Galileo, Newton, Maxwell, Faraday and the rest of us who are intrigued with how nature works had one thought in mind.

JJA: And that was?

Dr. Einstein: That is the quest to discover the "unifying" principle that explains nature. We seek a simple rather than a complex explanation. God's blueprint must be a simple pattern.

JJA: Why are you so confident of that?

Dr. Einstein: It is strange, but it is the one belief that bonds together everyone in the scientific club. It is, if I may say, a mystical belief that we cannot explain.

JJA: Dr. Einstein, may we explore another topic which may be quite sensitive to you and that is quantum mechanics? What is your objection?

Dr. Einstein: It is quite simple. I dislike not being able to measure things with precision—such as the location and/or speed of fundamental particles.

JJA: You believe, as did Newton, that everything in nature including electrons can be predicted with clocklike precision.

Dr. Einstein: Yes. This is an act of faith on my part, but I feel it in every synapse of my body. Man is uncertain; God is certain.

JJA: Hence, probability is another name for, "At this time, the explanation is unknown, but we have faith that eventually, we can and will know."

Dr. Einstein: Exactly. And we will be able to predict with mathematical precision for individual electrons and protons.

JJA: I have an alternate explanation I would like you to comment on, if you will. The experiments thus far support quantum mechanics. I'm thinking about the spin "up" and spin "down" demonstrations for electrons. Any individual electron may have a positive or negative spin and the probability is 50:50.

Dr. Einstein: Again, I believe that probability is an illusion, a surface mirage that does not represent the underlying reality. Let's take an example, not from physics but from medicine.

JJA: Good.

Dr. Einstein: I'm thinking of the British surgeon, Joseph Lister, who was appointed in 1861 as the head of the surgical wards at the Glasgow Royal Infirmary in Scotland. Here is probability: Lister noticed that if a patient had a wound that broke the skin or if he had to use his scalpel to cut a patient's skin, the chance that the patient would die after treatment was forty percent.

JJA: The uncertainty is: We did not know which group the individual patient would be in— the sixty percent who survived or the forty percent who perished.

Dr. Einstein: Precisely. There is your uncertainty principle.

JJA: As I recall, the hypothesis at the time was that wounds would infect as the result of exposure of the wound to oxygen. Hence, surgeons worked quickly to reduce exposure to a minimum.

Dr. Einstein: No one suspected that the surgeons themselves were spreading infections from one patient to another.

JJA: It did not occur to them that cleanliness was involved.

Dr. Einstein: Scalpels and other instruments were merely wiped on the surgeon's operating gown between operations. They kept a pitcher of water and a bowl nearby to clean the blood from their hands.

JJA: In 1865, Lister read about a recent article by a French chemist, Louis Pasteur. Pasteur observed through his microscope that microbes multiplied, producing decay.

Dr. Einstein: Could microbes be causing infections? That was a revolutionary hypothesis. Lister experimented with carbolic acid to kill minute particles in the air which he called "germs." It turned out that Lister's germ theory and antiseptic system of treatment was correct. The probability of surviving surgery went from 60 percent to 99 1/2 percent. But it was decades before surgeons accepted the "outrageous" germ theory.

JJA: Probability disappears when we discover the true cause and effect relationship.

Dr. Einstein: Ahhh. You are tuned in. But, you were about to suggest an alternative explanation.

JJA: It has to do with psychology.

Dr. Einstein: I don't believe there is such a thing as psychology or sociology.

JJA: Because of probability?

Dr. Einstein: Of course. After a hundred years of experiments by hundreds of talented researchers, there is no one equation that will predict the behavior of an individual human being.

JJA: Quite right. In fact, we can generalize further and say that there is not even one equation that will predict the behavior of any individual living creature.

Dr. Einstein: That includes animals, insects, microbes?

JJA: Yes. God, as you are fond of saying, does not play dice with the universe. Hence, we find equations everywhere in nature, except when we attempt to predict the behavior of individual living creatures.

Dr. Einstein: Hmmm. The behavior of individual living creatures is what? Mathematics free?

JJA: Exactly right. I believe the old testament calls it "free will."

Dr. Einstein: How does that apply to electrons?

JJA: What do you think?

Dr. Einstein: Are electrons mathematics-free? If so, then the implication is that electrons are living entities with intelligence.

JJA: Can it be? Let me ask you: Can electrons reproduce like microbes?

Dr. Einstein: Do you think that microbes are living entities?

JJA: They behave that way. They can reproduce and often guide their own genetic blueprint to mutate into other species. Even humans can't do that yet. Microbes can problem solve to overcome lethal agents such as penicillin that threaten them with extinction. That to me suggests intelligence.

Dr. Einstein: Me too. We still do not understand how microbes "adapt." How do they perform "plastic surgery" on themselves to transform into entirely different entities?

JJA: While I have your ear, let's explore the first law of thermodynamics.

Dr. Einstein: Heat moves in one direction from heat to cold but not the reverse.

JJA: I don't believe this is true.

Dr. Einstein: How so?

JJA: From observation of ice cubes, cold disappears at room temperature and becomes "non-cold" rather than hot. Heat does the same thing. Heat disappears and becomes "non-heat" rather than cold. There is symmetry.

Dr. Einstein: I agree. Let me think about that some more.

JJA: Now to a personal note: Perhaps no one is more responsible for the stereotype of the "absent-minded" professor than you.

Dr. Einstein: Perhaps that is true. I often called my wife to ask where I was, or directions to find my way home. But I was not the only one.

JJA: Whom did you have in mind, for instance.

Dr. Einstein: Sir Isaac Newton. He would often forget to eat and to sleep.

JJA: I understand that you would forget to eat and when you did eat, someone had to remind you to stop. Why so absorbed in thought?

Dr. Einstein: It is more exciting than anything else that is happening. We actually become those electrons and photons. We think the way they think. I call this fantasy, "thought experiments." But the truth is these mind trips are pure fun.

JJA: And when you discover something new on the trip?

Dr. Einstein: Elation. Pure elation. The journeys become an addiction. If something productive did *not* come out of it, you might say that we were on the edge of schizophrenia.

JJA: Then you do believe in psychology, at least abnormal psychology.

Dr. Einstein: I accept the descriptions of human behavior. I don't like depending upon probability to "explain" human behavior. I don't like the belief that God is uncertain in His design of the universe which is implied in the concept of "randomness." Even the theory of the "big bang" to start the universe bothers me.

JJA: The big bang is an implication from your theory of relativity.

Dr. Einstein: I know, but it still bothers me. It implies that the universe was set in motion by chance or randomness. When we understand the true cause and effect, randomness disappears.

JJA: It is strange that you use the name of God often but you were not a religious man.

Dr. Einstein: I did not practice religious rituals, but I was, as you call it, a "religious" man. How can anyone explore the secrets of nature for a lifetime and not see a marvelous craftsmanship in the exquisite designs? That craftsmanship is perfect and implies an intelligence that did not come from the human species. Exactly what the Intelligence is and how It works is the mystery, which perhaps will never be known.

JJA: You hesitated with your answer.

Dr. Einstein: Only because absolutes are so often off track completely. We believe we have a fact and the next generation demonstrates that it is a fiction.

JJA: Any example come to mind?

Dr. Einstein: The history of science and medicine are strewn with the intellectual corpses of "facts" that later were shown to be "fictions." One that comes to mind is Fulton's proposal for a railroad powered by a steam engine. Leading scientists of the day proclaimed that it would not work since it is a scientific "fact" that human beings cannot breathe at speeds of 30 miles per hour.

JJA: Dr. Einstein, the last decades of your life were a quest to find unification of relativity and quantum mechanics. Unification seems to be a scientific search for the Holy Grail.

Dr. Einstein: Yes. Newton's search for the Principia was a quest for the underlying principle that would unify gravity and electromagnetism. Descarte's search was to unify geometry and algebra which he did successfully in his immortal Analytic Geometry, reverently called, the Method.

JJA: Why this obsession with unification?

Dr. Einstein: I think it is a veiled search for the existence of God. It is an intellectual construction of that ancient biblical tower that will elevate us into God's residence somewhere in the sky. We believe, although no one will put it into words, that God's blueprint for the universe is elegant for its utter simplicity. Even Stephen Hawking talks about exploring the "Mind of God."

JJA: There is a book by Martin Rees entitled *Just Six Numbers: The Deep Forces That Shape the Universe*. In it, he presents six numbers which he says are critical to the creation of the universe.

His hypothesis is that by "chance" or by "accident," values for each number happen to fall within a narrow range that make the big bang for our universe possible. He prefers a "messy" physics in which the six numbers are disconnected to the aesthetic, "tidy" physics of Galileo, Descarte, Newton and yourself.

Dr. Einstein: It comes down to belief. Rees has an attractive idea which is that if there are billions of universes, by "chance" alone, one combination of circumstances will make for the creation of our universe with its own laws of physics. This is possible.

Dr. Einstein: But I prefer to believe that there are only two explanations for how the universe was created.

JJA: Only two?

Dr. Einstein: I think, only two. One is randomness which implies a chaotic coalition of unknown forces or gasses.

JJA: And the other?

Dr. Einstein: The other is the Pythagorean hypothesis that "The Good" which I like to call "The Old One," created the universe or universes with a Divine Blueprint that is clockwork perfect.

JJA: Universes?

Dr. Einstein: Yes, there may be billions of universes out there.

JJA: And "clockwork perfect" implies what?

Dr. Einstein: Patterns. Patterns. Patterns. Each pattern implies a cause-effect relationship. We are looking for cause-effect connections in nature. Every time a scientist or mathematician discovers a new pattern, there is excitement because once again we have evidence of handiwork that is not the product of a human mind.

JJA: How about an example of a new pattern from mathematics that anyone can understand?

Dr. Einstein: Sure. Take a look at this series:
$1 + 1/4 + 1/9 + 1/16 + 1/25 + 1/36 + 1/49 ...$

JJA: Well, let's see—we have:
$1 + 1/2^2 + 1/3^2 + 1/4^2 + 1/5^2 + 1/6^2 + 1/7^2...$
this continues forever. Now what?

Dr. Einstein: If you add these terms, the series converges to one number which is 1.6449.

JJA: So what is the mystery?

Dr. Einstein: The number 1.6449 was mysterious to 16th century mathematicians including such famous people as Leibniz, (who, independent of Sir Isaac Newton), discovered calculus, and the mathematical savants, Jakob and Johann Bernoulli.

JJA: Why is 1.6449 a mystery?

Dr. Einstein: It was unrecognizable.

JJA: Unrecognizable? I don't understand.

Dr. Einstein: It seemed to be a "random" number which contradicts the "pattern" view of the world. A series this basic with natural numbers should have a "pattern" in the divine blueprint.

JJA: The result of 1.6449 would be disturbing!

Dr. Einstein: It was. So, in 1734, Leonhard Euler (pronounced Oil-ler) decided to take a look. Euler was so successful in exploring mathematical mysteries that his writings filled hundreds of journal pages. But, the great Euler, too, was baffled.

JJA: How did he go about exploring this mystery?

Dr. Einstein: Like most mathematicians, he "doodled."

JJA: Doodled?

Dr. Einstein: Yes. He played with the puzzle like a child plays with a toy. He experimented with different numbers to test ideas that might give an explanation. That might yield a pattern.

But just before Euler was about to give up and admit defeat, he exclaimed, "...quite unexpectedly I have found an elegant formula..."

JJA: And he found?

Dr. Einstein: He found a hidden pattern. Everybody but Euler missed it. He discovered, just by playing with the numbers, that the series which equals 1.6449 is **pi** squared divided by six. In mathematical notation, it looks like this: $1.6449 = \pi^2/6$

JJA: 1.6449 is not a random number that appears by chance.

Dr. No, indeed. It is connected to **pi** which is a familiar concept known at least 2,000 years ago to the ancient Greeks.

JJA: We don't know why 1.6449 is connected to **pi** yet, but we have a new relationship—a new lead to explore.

Dr. Einstein: Exactly! The patterns are there, perhaps billions of them—invisible to the naked eye, waiting to be discovered. What could be more thrilling?

JJA: And anyone can play the game.

Dr. Einstein: Why not? It is only arithmetic... and I assure you that when you come closer and closer to discovering a new pattern, the Old One will not suddenly change the rules to trip you up.

JJA: If the game is so easy to play, why do you think we do not invite young school children to play?

Dr. Einstein: Marvelous question. The assumption is, I think, that they are incapable of playing the game. We underestimate them thereby condemning them to endure endless, repetitive drills.

JJA: Can you give me an example of a mathematical mystery that anyone - even school children - have a chance of solving?

Dr. Einstein: That's easy! Prime numbers.

JJA: But prime numbers have baffled eminent mathematicians for hundreds of years!

Dr. Einstein: But the problem is simple, and only involves arithmetic that is familiar to anyone with an elementary education.

JJA: Let's start at the beginning. What are prime numbers? Can you explain what they are in a way that any ordinary person can understand?

Dr. Einstein: Well, let me try with a few examples. Numbers are made up out of other numbers. For instance:
4 = 2 times 2
6 = 2 times 3
8 = 2 times 2 times 2
9 = 3 times 3
10 = 2 times 5

Notice that 2, 3, and 5 are "basic" or "prime" because there are no other numbers we can multiply together to create 2, 3, or 5.

JJA: And 2, 3, and 5 can be multiplied together to create other numbers. Some additional prime numbers are 7, 11, 13, 17, 19, and 23. So what is the mystery?

Dr. Einstein: First of all, *only one even number is prime.* That number is 2.

JJA: The second mystery is?

Dr. Einstein: Why are the remaining primes all odd numbers? And why did the Old One select some odd numbers to be prime, and not others? Mathematicians believe and so do I, that there must be a hidden pattern - a simple explanation, but what is it?

JJA: Why do you believe that there is an invisible pattern yet be discovered?

Dr. Einstein: Everything I've ever seen and all of my experience tells me that the universe is made of patterns. The Old One designed the universe with natural numbers such as 1, 2, 3...

Anything that fundamental has to - it simply has to have a pattern. Again, I don't believe the Old One let chance decide which numbers would be prime.

JJA: You believe that there is a simple equation that will produce every prime number?

Dr. Einstein: Yes. And anyone may discover it - even a school child - and become world famous!

JJA: Many mathematicians have been obsessed with this mystery, and have spent much of their working life attempting to find that elusive equation. How do you expect us to accomplish what experts have not been able to do for centuries?

Dr. Einstein: Why not? It's only arithmetic. Anyone can be a player!

JJA: Even a student in elementary school?

Dr. Einstein: Once a person has skill in addition, subtraction, multiplication, and division, they are ready to be a player. Now, can you suggest a way to start people thinking creatively about this mystery?

JJA: This is fun. Here we go. First, a simple layout of the natural numbers 1, 2, 3... into rows like this:

☐1	2	3	4	5	6
7	8	9	10	11	12
13	14	15	16	17	18
19	20	21	22	23	24
☐25	26	27	28	29	39
31	32	33	34	☐35	36
37	38	39	40	41	42
43	44	45	46	47	48
☐49	50	51	52	53	54
☐55	56	57	58	59	60
61	62	63	64	☐65	66
67	68	69	70	71	72
73	74	75	76	☐77	78
79	80	81	82	83	84
☐85	86	87	88	89	90
☐91	92	93	94	☐95	96
97	98	99	100	101	102

Dr. Einstein: I notice that *all the primes* are located in **column one** and in **column five**.

JJA: With some exceptions - which are in boxes. Those numbers in the boxes are false primes—that is odd numbers that are not prime.

Dr. Einstein: Another peculiar twist is that the numbers in **column one** and in **column five** can be predicted with these two simple equations:

$$6x + 1 \quad and \quad 6x + 5$$

JJA: These equations may produce every prime number in *column one* and *column five* :
6x + 1 and 6x + 5

	6x + 1	6x + 5
0	$\boxed{6(0) + 1 = 1}$	6(0) + 5 = 5
1	6(1) + 1 = 7	6(1) + 5 = 11
2	6(2) + 1 = 13	6(2) + 5 = 17
3	6(3) + 1 = 19	6(3) + 5 = 23
4	$\boxed{6(4) + 1 = 25}$	6(4) + 5 = 29
5	6(5) + 1 = 31	$\boxed{6(5) + 5 = 35}$
6	6(6) + 1 = 37	6(6) + 5 = 41
7	6(7) + 1 = 43	6(7) + 5 = 47
8	$\boxed{6(8) + 1 = 49}$	6(8) + 5 = 53
9	$\boxed{6(9) + 1 = 55}$	6(9) + 5 = 59
10	6(10) + 1 = 61	$\boxed{6(10) + 5 = 65}$

Dr. Einstein: But we have a glitch! We are picking up some "false primes" - enclosed by the boxes above.

JJA: We can find some odd numbers that are non-prime with: 6x + 9. The challenge: Discover another equation (currently hidden from our view), to produce the remaining odd numbers that are non-prime.

	6x + 9	(Mystery equation?)
0	6(0) + 9 = 9	= 25
1	6(1) + 9 = 15	= 35
2	6(2) + 9 = 21	= 49
3	6(3) + 9 = 27	= 55
4	6(4) + 9 = 33	= 65
5	6(5) + 9 = 39	= 75
6	6(6) + 9 = 45	= 77
7	6(7) + 9 = 51	= 81
8	6(8) + 9 = 57	= 85
9	6(9) + 9 = 63	= 91
10	6(10) + 9 = 69	= 93

Dr. Einstein: For symmetry, there should be an equation that will give us the remaining odd numbers that are non-prime or false prime. I would have the children looking for the missing mystery equation.

JJA: Why are you so confident that the mystery equation exists?

Dr. Einstein: Because I have observed so many intricate patterns in nature—from the microscopic to the gigantic movement of planets—patterns that are amazingly predictable with mathematics.

I believe that God is not "messy." The Old One did not reach into the inside pocket of His immaculate white cloak for dice to decide which odd numbers would be prime. He is certain in His design which is clean, simple, and symmetrical.

As the children search for the mystery equation using simple arithmetic, who knows? One of the youngsters may find that *golden equation*—the one that will produce *all primes* and no false primes. It's there. It has to be.

JJA: A golden equation. Another possibility: How about a silver equation—one that will produce *some primes* and no false primes?

Dr. Einstein: Why not? And, let's add a bronze equation that will produce all false primes (odd numbers that are non-prime). Any of those discoveries will make someone world famous in the world of mathematics and science—and that someone could be a student in elementary or middle school. This is the romance of mathematics!

JJA: It seems to me that we are close to a golden equation with 1.2x - 1 which for some unknown reason produces impressive results when we only look at outcomes that are a whole number or end in .4:

2	$\boxed{1.2\,(2) - 1 = 1.4}$	52	$1.2\,(52) - 1 = \mathbf{61.4}$
5	$1.2\,(5) - 1 = \mathbf{5}$	55	$\boxed{1.2\,(55) - 1 = 65}$
7	$1.2\,(7) - 1 = \mathbf{7.4}$	57	$1.2\,(57) - 1 = \mathbf{67}$
10	$1.2\,(10) - 1 = \mathbf{11}$	60	$1.2\,(60) - 1 = \mathbf{71}$
12	$1.2\,(12) - 1 = \mathbf{13.4}$	62	$1.2\,(62) - 1 = \mathbf{73.4}$
15	$1.2\,(15) - 1 = \mathbf{17}$	65	$\boxed{1.2\,(65) - 1 = 77}$
17	$1.2\,(17) - 1 = \mathbf{19.4}$	67	$1.2\,(67) - 1 = \mathbf{79.4}$
20	$1.2\,(20) - 1 = \mathbf{23}$	70	$1.2\,(70) - 1 = \mathbf{83}$
22	$\boxed{1.2\,(22) - 1 = 25.4}$	75	$1.2\,(75) - 1 = \mathbf{89}$
25	$1.2\,(25) - 1 = \mathbf{29}$	77	$\boxed{1.2\,(77) - 1 = 91.4}$
27	$1.2\,(27) - 1 = \mathbf{31}$	80	$\boxed{1.2\,(80) - 1 = 95}$
30	$\boxed{1.2\,(30) - 1 = 35}$	82	$1.2\,(82) - 1 = \mathbf{97.4}$
32	$1.2\,(32) - 1 = \mathbf{37}$	85	$1.2\,(85) - 1 = \mathbf{101}$
35	$1.2\,(35) - 1 = \mathbf{41}$	87	$1.2\,(87) - 1 = \mathbf{103.4}$
37	$1.2\,(37) - 1 = \mathbf{43.4}$	90	$1.2\,(90) - 1 = \mathbf{107}$
40	$1.2\,(40) - 1 = \mathbf{47}$	92	$1.2\,(92) - 1 = \mathbf{109.4}$
42	$\boxed{1.2\,(42) - 1 = 49}$	95	$1.2\,(95) - 1 = \mathbf{113}$
45	$1.2\,(45) - 1 = \mathbf{53}$	97	$\boxed{1.2\,(97) - 1 = 115.4}$
47	$\boxed{1.2\,(47) - 1 = 55.4}$	100	$\boxed{1.2\,(100) - 1 = 119}$
50	$1.2\,(50) - 1 = \mathbf{59}$		

Dr. Einstein: Not bad. It is a quasi-golden equation because it gives us almost every one of the 30 primes located between the counting numbers of 1 and 100. The primes of 2 and 3 are missing and....

JJA: But we still have eleven *false primes* which are in the boxes above.

Dr. Einstein: Yes. The target for people searching for a better equation is to generate at least the 28 primes located between the numbers 5 and 100 and yet have less than eleven false primes—preferably zero false primes. If the new equation works for the set of natural numbers between 5 and 100, it will probably work for numbers between 101 and 200, and so on.

JJA: Again and again I hear you refer to God or The Old One. You are said to be an atheist, yet no one in science used the word, God (or as you sometimes said, The Old One) more frequently than you.

Dr. Einstein: I admit it. I am a bundle of contradictions. I was a pacifist and anti-militarist all my life but with my endorsement, the President of the United States dedicated enormous resources to develop the atomic bomb. I was against nationalism but I supported the state of Israel. I loved humanity but I was close to very few people.

JJA: You call yourself an atheist, but in 1934 you were quoted as saying, " The subject of psychic research deserves serious consideration."

Dr. Einstein: There is abundant evidence that there is something to parapsychology. At this time, it is not as predictable as the elliptical path of the earth around the sun. Hence, more work is necessary to discover why a few people seem to have some unusual gifts such as telepathy.

Perhaps we all have these gifts but they are dormant. What is the mechanism? How does it work? How can we call upon it on cue?

JJA: Any specific suggestions for experimental verification of psychic phenomenon?

Dr. Einstein: I have one suggestion. If some people have the ability to communicate with those who have passed into another dimension, then try to make contact with mathematicians and scientists. Ask for a specific answer to a specific question that can be tested with experiments. Why not? I can't think of a more powerful format for verifying communication with the departed, can you?

JJA: Interesting idea. I cannot close without asking you about the invitation in 1934 to join President Franklin D. Roosevelt and Eleanor for dinner at the White House.

Dr. Einstein: It was a delightful dinner with convivial people. My wife and I not only had dinner with the President and his wife but we stayed overnight.

JJA: What impressed you about FDR?

Dr. Einstein: His wit, charm, and intellect. I was especially surprised at how fluent he was in German.

Summary

My conversation with Dr. Einstein is free-wheeling. We explore everything from his belief in nature's grand design to the existence of God. As we get closer to answering life's riddles, he believes that God will not suddenly change the rules to trip us up.

I asked him about psychology and sociology. Since these disciplines are built on probability statements, are they credible? His answer surprised me.

CHAPTER 12

Conversation with Sir Isaac Newton
about a new kind of mathematics,
the secret practice of the illegal "black arts,"
and his fascination with the Bible.

JJA: Sir Isaac Newton, whoever coined the word "Scholar" with a large "S" had you in mind. No one in history matches your dedication, insight, meticulous attention to detail, technical skills, and work ethic. Your book, *Philosophiæ Naturalis Principia Mathematica* (Mathematical Principles of Natural Philosophy) revolutionized the way the universe is perceived. Yet, Sir Isaac, the first edition of the book sold only a few hundred copies within a decade of its publication in 1687. Why is that?

Sir Isaac Newton: I think the ideas were too novel for the scientific mind set at that time. Later, the book enjoyed more than 100 editions and has been translated and printed in every language on earth.

JJA: Did you expect that kind of colossal impact?

Sir Isaac Newton: No. First, I must say that the book was written as a work in progress, not a final picture of the universe. And second, I tried to make the book as unreadable as possible.

JJA: May I ask why? I know that you prefer people not probe into your life. Please let me know if I transgress.

Sir Isaac Newton: You can be sure of that. To answer your question, I made the book unreadable to avoid being baited by the little "smatterers" of mathematics.

JJA: Smatterers?

Sir Isaac Newton: Yes, I like the term "smatterers." Some people would call them dabblers.

JJA: Speaking of mathematics, you are remembered as one of the greatest mathematicians in history, yet I understand that when you entered Cambridge at the age of nineteen, your mathematical skill was limited almost completely to arithmetic. Is that accurate?

Sir Isaac Newton: Yes. I taught myself other mathematical skills from books as I needed to know about things. For example, when I began to explore astronomy, I realized I needed to understand trigonometry—so I bought a book on the subject.

JJA: When you entered the university, you were a sizare. Can you explain the duties of a sizare?

Sir Isaac Newton: I supplemented my income be emptying bed pans and delivering meals to wealthy upper class students—that's what a sizare does at Cambridge.

JJA: Can you describe the village of Cambridge in those days?

Sir Isaac Newton: It was more like a squatters camp than a village. It was a dirty, squalid, cramped place with streets so narrow that only one wheelbarrow could navigate at a time. You must remember that in England of the 17th century, 90% of the inhabitants were living in utter poverty while only 10% enjoyed the good life as ladies and gentlemen.

JJA: What is there in *Principia* that revolutionized the way we perceive the universe?

Sir Isaac Newton: The book explains how the tides are produced, how comets travel through the heavens, and why the earth "wobbles." With my calculus, I checked the validity of Kepler's third law by finding the velocity of planets at different points on their elliptical path.

JJA: How did you feel when you discovered that the mathematical calculations tallied precisely with observed facts about planetary orbits?

Sir Isaac Newton: I was shocked. The excitement of these discoveries would keep me working for days without food or sleep.

JJA: The picture we have is of you sitting in the garden of Trinity College at Cambridge all night, wrapped in blankets, observing planetary movements.

Sir Isaac Newton: Sleep deprivation produced a kind of psychosis. I was woozy and found myself writing skewed letters to my friends, which baffled them. When I regained my balance, I was embarrassed and wrote letters of apology.

JJA: Sir Isaac, your calculus fascinates me not only as a stunning discovery with a lovely symmetry, like addition and subtraction or multiplication and division, but there are other more exotic implications.

Sir Isaac Newton: Like what?

JJA: I hesitate to say it, but your secretive experiments with alchemy are now well documented. You wrote over a million words about the occult and alchemy.

Sir Isaac Newton: Of course I was secretive. It was a capital offense in England of my day to explore alchemy. Yes, I was secretive and fearful. Wouldn't you be?

JJA: Yes, indeed. Why do you think that a practitioner of the art risked execution?

Sir Isaac Newton: Alchemy is the quest to turn something of less value into something of more value like turning lead or other base metals into gold. If this secret were to be discovered, imagine how it would upset the social order.

If anyone could produce silver or gold, then the privileged classes would disappear. Everyone would be the leisure class and no one would be available to work!

JJA: What strikes me about your calculus is the analogy with alchemy.

Sir Isaac Newton: How so?

JJA: You led a double life—to the world you were a scientist and mathematician but you had a secluded, hidden laboratory with a furnace and chemicals. You experimented with potions and recipes of alchemy. You were a very serious practitioner of the black arts.

Sir Isaac Newton: I was a serious person. My assistant commented that he only heard me laugh once in all the years we worked together. I don't know—is that a compliment?

JJA: It must have been a pretty good joke to make you laugh! Well, I don't mean to sound disrespectful. Your dedication certainly paid handsome dividends. Your were knighted by Queen Anne and declared a national treasure. Your discoveries changed the world forever.

Sir Isaac Newton: Thank you. I am curious. What connection do you see between alchemy and calculus?

JJA: In calculus, you take a formula of less value and convert it into another formula of more value. Then that new formula is converted again into still another formula of high value. It is like taking lead and converting it into silver, then taking the silver and converting it into gold.

Sir Isaac Newton: The analogy is a good one. But there must be other examples of this in mathematics.

JJA: Maybe not. I notice in statistics that, with a nimble use of algebra, one procedure can be transformed into another and still another to produce the *t test*, *chi square*, *correlation*, and *analysis of variance*. But all these procedures are equivalent. They appear to be different, but all are answering the same question. Not so with calculus. You truly did find the secret for transforming mathematical "base metals" into "silver and gold."

Sir Isaac Newton: I was thrilled to discover that the alchemist principle had validity. Calculus does not produce something from nothing, but it does start with a formula and produce another formula with more information than the initial formula and then with another transformation, we get even more information. With each transformation, the formulas answer a different question about nature

JJA: You acknowledge the role of alchemy in your work.

Sir Isaac Newton: Yes. My life was an alchemist's quest to find the simple, unifying principle of nature. I believe it is there, invisible, but waiting to be seen.

JJA: Unifying principle. Please explain your thinking about this.

Sir Isaac Newton: It is sometimes called the philosopher's stone. I believe that there is an underlying principle that explains all forces in nature such as gravity, magnetism, and electricity. I believe, that one principle, a simple concept, accounts for microscopic as well as telescopic phenomena

JJA: Even today, scientists around the world are seeking the philosopher's stone. Do you think we are genetically programmed with a primeval awareness of a simple principle explaining all the forces of nature?

Sir Isaac Newton: Yes, I firmly believe this is the case.

JJA: "Simplicity" seems to be a key word in your philosophy. Your life was focused upon finding a simple explanation for nature that was hidden from view. You lived an austere, almost monastic life style. In the spirit of "simplicity," can you explain to me the principle of calculus so that an elementary school child can understand it?

Sir Isaac Newton: I was unable to explain anything that I discovered with enough simplicity that Cambridge students could understand. Perhaps that is why nobody attended my lectures.

JJA: You spoke to an empty room?

Sir Isaac Newton: Almost always. If the lecture was scheduled for thirty minutes and no one was in attendance, I would speak to a wall across the room for fifteen minutes, then retire to my laboratory to conduct more experiments. Teaching was not one of my skills.

JJA: Then, I am asking too much perhaps...

Sir Isaac Newton: No, no. I'll give it a try.

JJA: Thank you.

Sir Isaac Newton: Calculus is finding information about curved lines on a graph. For example, how fast is something moving along that curved line?

JJA: Calculus is finding how fast something is moving along a curved line. Why is this revolutionary?

Sir Isaac Newton: Before calculus, we could only determine velocity for straight lines on a graph. If the line on a graph curved, no one knew what to do.

JJA: Why is this important?

Sir Isaac Newton: Objects in space change their velocity and if we plot this on paper, we get curves rather than straight lines.

JJA: And the problem is we cannot measure a "curve"...

Sir Isaac Newton: That's right. No one has discovered a way to measure a curve, but if we could "zoom in" on the curve with a powerful enough microscope, we wouldn't see a "curve" — we would see tiny straight lines. The concept of a "curve" may be an illusion.

JJA: And tiny straight lines can be measured with simple geometry.

Sir Isaac Newton: We can, for example find the slope at any point on the curve, and the area under any segment of the curve. That's the essence of calculus.

JJA: Sir Isaac, the most enduring story about you has to do with an apple falling from a tree. Michael White, one of your biographers, calls the story a myth perhaps created by you to hide your secret alchemy experiments.

Sir Isaac Newton: Why a myth?

JJA: The falling apple implies that your model of gravity was a flash of inspiration.

Sir Isaac Newton: There was indeed inspiration, but chance works best with a prepared mind. I worked endless hours on the problem and my knowledge of... well I can say it now... the "black arts" such as alchemy skewed my thinking enough to trigger exciting insights.

JJA: How did brainswitching fit your style of discovery?

Sir Isaac Newton: Brainswitching? Please clarify.

JJA: This is working the right and left sides of the brain to solve problems.

Sir Isaac Newton: The concept of right and left brain is alien to me.

JJA: One of your lifelong habits was to write down questions in your notebooks. You called them "queries" about topics that intrigued you. For example, Query 23 for Opticks reads, "By what means do bodies act on one another at a distance?"

Sir Isaac Newton: Yes. I found this exercise to be invaluable. Mysteriously, answers would later come to me and I would note those down in my notebook.

JJA: That is a marvelous example of brainswitching. The left brain, the verbal side, formulates the question and submits it in writing to the right brain, the mute, creative side for an answer.

The right side processes the question and recombines bits and pieces of information until it has an "answer." If the left brain evaluates the "answer" as "silly" or "unrealistic," then the right brain will continue to search for alternate possibilities until you respond with, "Ahhh. That's it!"

Sir Isaac Newton: I did that all my life.

JJA: I know. You were a master at the art of brainswitching.

Sir Isaac Newton: The question and answer ritual gave me a private feeling that I was in communication with a divine force—the universal mind that contains all knowledge. All I had to do was ask.

JJA: And from the Bible: "Ask and ye shall receive, knock and the door shall be opened."

Sir Isaac Newton: As you probably know, I explored the Bible in great detail to find answers for myself to spiritual questions.

JJA: For instance?

Sir Isaac Newton: Well, I could not find a single reference to the soul. The word "soul" is nowhere in the Bible.

JJA: You were particularly fascinated with the Book of Prophecy and the Book of Genesis.

Sir Isaac Newton: I concluded that I believe the universe was created in six days.

JJA: What was your reasoning?

Sir Isaac Newton: The length of a day is not specified. Hence, a "biblical day" is not necessarily twenty-four hours. God used two "days" to create the planetary system before creating the earth. Each "biblical day" could have been millions if not billions of years.

JJA: You had problems accepting the concept of a trinity.

Sir Isaac Newton: Yes, my mathematical instincts did not permit me to accept that 3 is equal to 1 or 1 is equal to 3.

JJA: It is curious that in England of the 17th century, there was religious tolerance except for Catholics and Aryans, of which you were one. What do Aryans believe?

Sir Isaac Newton: We believe that there are special people on earth who are destined to make great contributions to the human race.

JJA: What about God? What do Aryans think?

Sir Isaac Newton: They believe that God does nothing by Himself which He can do through others. Jesus Christ was the "first creature" who was neither of the same substance as God nor human. He is special. He is a medium through which God's will is done.

JJA: On your deathbed, you refused the last rites which included confession.

Sir Isaac Newton: I had nothing to confess. As a young man, I faithfully kept a notebook full of confessions which unburdened me of sin.

JJA: Sir Isaac, no offense, but your entire life was one of duplicity.

Sir Isaac Newton: Oh! How so?

JJA: Although you zealously explored the Bible for spiritual answers you were equally obsessed with the antithesis of the Bible, the black arts.

Sir Isaac Newton: I wrote in my notebooks more than a million words about alchemy and the occult. But what makes you think these are antithetical? All are part of the Divine Blueprint. Underlying everything is a simplicity and harmony that may never be known... it may be an eternal mystery that is unknowable.

JJA: What is your personal view of truth? You once wrote in one of your notebooks: "Plato is my friend... but truth is my *best* friend."

Sir Isaac Newton: Looking back on my life, I was like a child playing on the beach surrounded by an ocean of mystery searching for a stone smoother than the others and a sea shell prettier than the others.

Summary

My conversation with the amazing Sir Isaac Newton revealed a secret life—full of contractions. He explored the Bible for esoteric answers with the same intensity as he experimented with the occult and alchemy. He was curious about everything.

He wrote his thoughts and questions in hundreds of private notebooks. If he did not know something, he acquired books and taught himself. He imposed no self-limits. He never said to himself, "I can't do that!"

When telescopes at the time gave only fuzzy views of planets, he set out to built his own telescope with precision lenses that gave a clear, sharp view of cosmic events. Reluctantly, he displayed his telescope at the Royal Academy in London. The members were dazzled by the exquisite craftsmanship of the instrument.

CHAPTER 13

A conversation with Dr. Richard P. Feynman, Nobel Laureate in Physics.

The Strange Mystery of Light, Quantum Mechanics, and Zeno's Paradox.

JJA: Dr. Feynman, What a pleasure to talk with you. You are every student's hero.

Dr. Feynman: Thank you. To be a hero of students is almost better than winning a Nobel prize. But, why am I a hero?

JJA: Two reasons: First, you talked *to* students rather than talking *at* them.

Dr. Feynman: And second?

JJA: Second, you were able to "read their minds."

Dr. Feynman: Really? I was a mind reader. How do you figure?

JJA: You were able to anticipate how that critical left brain of students would attempt to derail them with messages such as: "I don't understand what he's saying! I don't get it! I must be the only person here who is baffled. I guess I'm no good as physics."

Dr. Feynman: Well, you may have something there...

JJA: I know I do. For example, you got students in the right mood for listening to you by telling them that "There are many reasons you will not understand what I am about to explain. That's OK because... and then you held up all of the red flags that come up in the student's head so that you quieted that critical left brain—letting their right brain get access to your message and enjoy what they are hearing.

Dr. Feynman: I guess I did what any crackerjack of a salesman does.

JJA: On page 24 of your classic book, QED,* you told your audience: "Brace yourself for this—not because it is difficult to understand, but because it is absolutely ridiculous." If you were my physic's teacher in high school, nothing would block me from pursuing a career in physics. To me you are the Will Rogers of physics—playfully spinning your lariat as you explore the mysteries of light.

Dr. Feynman: My Stetson pushed back from my forehead as I move around the stage in my Levis and scuffed-up cowboy boots... not bad for a kid from New York City.

JJA: You often said that one of the perks of winning a Nobel prize is that you get to ask dumb questions. People figure that even though the question is stupid, it must have some hidden significance since you are a Nobel prize winner.

Dr. Feynman: I think my entire life has been one of asking dumb questions. They are the most profound, but they disturb people because even the experts

---

* Feynman, Richard P. ***QED: The Strange Theory of Light and Matter*** 1985. Princeton University Press., 41 William Street., Princeton , NJ 08540

don't know the answers. And nobody wants **not** to know the answers, especially professionals in any occupation. So, they make it look as if you are foolish for asking.

JJA: Speaking of "dumb" questions, do you know what happened to Thomas Edison?

Dr. Feynman: No, what?

JJA: He entered school at the age of 7 but his mother withdrew him after three months and he never returned to school.

Dr. Feynman: Why?

JJA: Because the teacher continually whipped him for asking so many questions.

Dr. Feynman: Were they really dumb questions?

JJA: The schoolmaster must have thought so because he was labeled as "addled." He was curious about everything. He wanted to know: "Why do eggs sizzle in the frying pan?" "Why do birds fly?" "Why does water put out fire?" and "How does a hen hatch chicks?" When no one answered his questions, he would experiment, trying to get answers.

Dr. Feynman: I remember that he collected some eggs and sat on them to see whether heat from his body would hatch the eggs. Tell me, with only three months of schooling, where did he get his primary education?

JJA: His mother pioneered "home schooling." She was an innovative school teacher herself and made a game of teaching him which she called "exploring the exciting world of knowledge." Learning was pure fun. The boy learned so fast that his mother could no longer teach him.

Dr. Feynman: I knew there was a reason I liked Edison so much.

JJA: He reminds me of you, especially when Mrs. Edison gave the 9 year old boy a chemistry book. Edison would not accept as true the statements made in the book and proceeded to test every experiment himself to prove the author wrong.

Dr. Feynman: Was I that cantankerous?

JJA: No, just curious— much like Einstein.

Dr. Feynman: Einstein once remarked, " I have no particular talent except that I am exceedingly inquisitive."

JJA: The chemistry book story reminds me of your early days. I was impressed with your approach to algebra and trigonometry.

Dr. Feynman: What was so special?

JJA: You did what most kids would not do. You went to the library and checked out some books. Most kids would think: "I can't figure that out on my own. I will do like everyone else—wait, and take a class in school."

Dr. Feynman: Well, I was fascinated with science and I realized that I needed a box of mathematical tools to explore the mysteries of science.

JJA: So you worked all the exercises in the algebra book and then tried to "prove" the theorems on your own.

Dr. Feynman: Yep! I discovered later that sometimes I was right on target; sometimes I was way off-target and—amazingly, sometimes, my proof was better than the book.

JJA: What motivated you to undertake learning algebra on your own? Must people would tell themselves: "It's impossible! I can't do it!"

Dr. Feynman: I am weird. I admit it. I always told myself, "How difficult can it be?" And I had a compelling reason to want to learn.

JJA: One of my colleagues once remarked: " You can't learn to program a computer unless you have something you want to program." You could foresee applications of algebra in your life.

Dr. Feynman: Yes.

JJA: Once you finished algebra, you opened a book of trigonometry. Again, you jumped out of the box.

Dr. Feynman: What do you mean?

JJA: Most people would be intimidated with the concepts, but you proceeded exercise by exercise and then did an extraordinary thing.

Dr. Feynman: What's that?

JJA: You decided that the names of things in the world of triangles were meaningless to you—so you made up your own language.

Dr. Feynman: And it worked for me. Everything seemed to fit together—it continued to work for me almost through graduate school until I began to work with other professionals and realized that they did not know what I was talking about. I must have sounded schizophrenic.

JJA; Again, I am impressed. Most people would be scared to death to deviate in any way from what is in a math book.

Dr. Feynman: Yeah. That is indeed a mistake. Unless you translate that gobbledygook into your language, you will never understand what it is all about.

JJA: That thought should be engraved in gold leaf and hung in every math classroom.

Dr. Feynman: Amen!

JJA: I wanted to explore several mysteries with you. One has to do with the nature of light. Most people seem to believe that light has the dual nature of being sometimes particles and sometimes waves.

Dr. Feynman: Absolutely a myth. Light is particles, period.

JJA: You convinced me that light is particles in your sensational book, QED. If I had my way, that would be the first book physics students ever read. Simple, clear, and immediately understandable. It would hook every kid on physics.

Dr. Feynman: Now that I think about it, that's a marvelous idea!

JJA: You come in conflict with Dr. Einstein—not that he is the ultimate authority—but your conclusion about probability of photon behavior would be disturbing to Einstein.

Dr. Feynman: I know it. It disturbs me. It disturbs all the physicists. We like dealing in exactitude, not probabilities. But, unfortunately, nature may have different ideas.

JJA: More specifically, you state that if 100 photons from light enter a plate of glass, only about 4 will be reflected from the glass—and here is the mystery. We don't know from experiment to experiment, which 4 photons in the 100 will emerge.

Dr. Feynman: That is the uncertainty principle in Quantum Electrodynamics—or you may know it as Quantum Mechanics.

JJA: I'm curious. How are you able to track each individual photon in the set of 100? How do you know the identity of each? If they were animals in the wild, you might tranquilize the critter and attach a miniature radio transmitter to an ear.

Dr. Feynman: Whoa! That's one of those penetrating questions which is the heart of QED. How do we know the identity of each photon? Here's the answer—better yet, let me demonstrate that your right brain already knows the answer.

JJA: It does?

Dr. Feynman: Yes.

JJA: Please proceed.

Dr. Feynman: You're going on a guided journey. I will be the train conductor. Let's shine a light of the same color, say red, on a sheet of glass that has a specific thickness. The first time we conduct the experiment, of 100 photons, how many will bounce off the surface?

JJA: Well, let's see... you mentioned that on the average, we expect 4 in 100 to reflect from the glass' surface.

Dr. Feynman: So, this is Experiment Number 1. How many clicks will we hear on the photomultiplier?

JJA: It could be 1, 2, 3, 4, 5, 6, or 7

Dr. Feynman: How do you figure?

JJA: Since the average is four percent, the clicks from experiment to experiment could vary from 1 to 7 because the sum of $1 + 2 + 3 + 4 + 5 + 6 + 7$ equals 28. Divide 28 by 7 to get an average number of clicks which is 4.

Dr. Feynman: When we repeat the experiment over and over, we will hear four clicks more often than 1, 2, 3, or 5, 6, 7. Now, does the answer to your question jump out at you?

JJA: I think so. If we got 4 clicks every time we turned on the light, then perhaps we could make a case that the same 4 photons were bouncing off the glass.

Dr. Feynman: But, since the clicks could be any number from 1 to 7...

JJA: We don't know why sometimes only one photon "decided" to reflect from the surface of the glass and the other 99 "decided" to continue on through the glass and...

Dr. Feynman: ...and other times, two photons or three or four or more perceived that the surface of the glass is a trampoline.

JJA: Hence, we don't know exactly which photons in 100 will bounce off the trampoline.

Dr. Feynman: It would be helpful if we could monitor each photon like dots on a radar screen because then we could plot the trajectory of each to solve this deep mystery. But the technology does not exist today.

JJA: If photons are making decisions, are you implying that they have intelligence?

Dr. Feynman: I don't think so, but anything is possible.

JJA: In any event, we are now into probability, which really disturbed Dr. Einstein.

Dr. Feynman: Yep! Nobody likes the uncertainty in QED, but that is the hand that nature dealt us to play. Now what do we do with it?

JJA: You know from my conversation with Dr. Einstein that we played with the notion that electrons may be biological. Are photons able to reproduce or is that a dumb, dumb question?

Dr. Feynman: A double dumb question. You know that I love dumb questions. I love double dumb questions even more.

My answer is that I don't know. I don't think so, but no one really knows.

Incidentally, it is important for a teacher to editorialize when it is appropriate by telling students: "I don't know why this is so. Nobody knows why this is so." Otherwise, students will blame themselves with: "I don't get it! I don't understand! Am I the only one in the room who is baffled?"

JJA: Good advice. Dr. Feynman, tell me, can an experiment be created to test the hypothesis that photons are capable of reproducing themselves?

Dr. Feynman: Sure. Why not?

JJA: When you use probability in describing the behavior of photons, Einstein would say: "You don't know exactly what is happening in the black box, so you pretend to know with a probability statement."

Dr. Feynman: Well, he is right and he is not. We can at least predict how photons will behave as a group. For example, from 100 photons going into the glass, about 4 will be reflected, but we cannot predict which 4. This seats physicists at the table with the psychologists and sociologists, playing their probability game. That's quite a shift.

JJA: If each individual photon is unpredictable, then is it possible that they have both intelligence and "freewill"? If so, then the individual photon is mathematics-free—there is no equation that will predict it's behavior.

Dr. Feynman: You explored this delicate issue with Dr. Einstein and yes, it is possible. Entities in physics and chemistry may indeed be biological. That is a mystery yet to be explored.

JJA: When we draw boxes around each discipline, we tend to think that each is independent of the other.

Dr. Feynman: An illusion, to be sure. I feel, for instance, that chemistry and physics are one discipline. They overlap, and if we knew enough, they would merge into one, unified body of knowledge. We can throw mathematics in there too. When a new mathematic is discovered, it seems to be "useless." That is, no practical application is apparent at the time. But later, someone finds an exciting connection to something vital.

JJA: How about an example?

Dr. Feynman: During the late 1600's, the German mathematician Gottfried Wilhelm Leibniz played with a system of numbers using only 0 and 1. It had no practical application at the time. It was a curio which only a handful of mathematicians appreciated.

About 250 years later, in the 1940's, we discovered that Leibniz' strange 0 and 1 number system could be used in machines that translated text and computations at lightening speed. Those machines are now known as computers.

JJA: Ahhh. You are a Pythagorean at heart. You feel that there is some grand design—some unifying principle in the universe.

Dr. Feynman: Absolutely. My personal metronome is synchronized with the Newton's clockwork universe which came from Pythagoras.

JJA: Dr. Feynman, you mentioned in QED that there are no half-photons. It is all or nothing.

Dr. Feynman: Yes.

JJA: I would like to explore one of Zeno's paradoxes with you. It is connected with the notion that there are no half-photons.

Dr. Feynman: The mystery of infinity. Let's look at it... Did you discovery who won the race—the hare or the tortoise?

JJA: Curiously, I think I know which one won the race.

Dr. Feynman: Well, let's see... Zeno, as I recall, places the hare at the starting mark and the tortoise one step ahead, then fires a starter pistol. If the hare must go a *half a step* before reaching the tortoise and then *another half* of the *half-step* before that and so on, the hare will never reach the tortoise.

JJA: Right! The hare is trapped in endless backward motion into infinity that almost, but never reaches zero, the starting position for the hare.

Dr. Feynman: The tortoise moves off the first step and he also begins an endless regression into infinity so that he will never reach the second step.

JJA: The hare never reaches the first step and the tortoise never reaches the second step. Hence, the hare never overtakes his slow-moving competitor.

Dr. Feynman: It looks as if neither one can win the race. The hare is trapped between zero and step one while the tortoise is trapped between step one and step two.

JJA: I think that the half-step concept in Zeno's race, is analogous to a half-photon, which you stated, does not exist. I don't think Zeno's half-step exists either.

De. Feynman: How so?

JJA: I think it works like this: If the distance the two must run is two steps, then the instant the starter pistol is fired, the hare moves to step one and the tortoise moves to step two which is the goal line and wins the race.

Step 1	Step 2

Dr. Feynman: What about Zeno's half-step?

JJA: There is no half-step. If you want a half-step, the distance to the goal instantly changes from two steps to four steps.

Step 1	Step 2	Step 3	Step 4

The starter pistol goes "bang." The hare moves to step one and the tortoise moves to step three. The next instance, the hare moves to step two and the tortoise moves to step four and wins the race.

Dr. Feynman: Since there are no half-steps, if we decide that we want half-steps, the scale will change again before the race begins. It will change from four steps to eight steps.

Step 1	Step 2	Step 3	Step 4	Step 5	Step 6	Step 7	Step 8

JJA: Yes. "Bang!" The hare moves to step one and the tortoise moves to step five. In the next instance, the hare moves to step two and the tortoise moves to step six. We assume, as did Zeno, that both creatures are moving at the same velocity— one step at each instant. The hare will never overtake the tortoise to win the race.

Dr. Feynman: Conclusion. Steps within a scale are discrete. They are finite, but the scale can alter the number of steps within its boundaries forever. The scales are infinite but the steps within each scale are finite.

JJA: Now, back to those photons coming from the light. I think that eventually Dr. Einstein will prove to be right. When we understand the true cause and effect, we will be able to predict with exactitude why a different set of four photons bounce off the glass each time the light is turned on.

Dr. Feynman: Entirely possible. My instincts tell me it may turn out like that. For the moment, we seem to be stuck with probability.

Summary

Of all the sciences, physics seems to be the most exact, but is it? This may be an illusion, as I discover in my conversation with the 1965 Nobel prize winner, Dr. Richard P. Feynman. Yet, he too is a fan of Pythagoras and a believer in the precise, clockwork universe of Newton. He explains the mystery of light with such simplicity that everyone can understand. His passion for physics will capture you and give you the feeling that, "Not only do I understand, but I think I can solve this mystery!"

CHAPTER 14
Another Conversation with Dr. Albert Einstein

Do primes* exist... or are they the greatest mathematical myth in history?

JJA: Welcome, Dr. Einstein. I appreciate your return to explore the fascinating concept of primes which, as you know, are fundamental to arithmetic.

Dr. Einstein: Fundamental is the key word. If you intend to challenge "primes," you are in for intellectual guerilla warfare from mathematicians.

JJA: Why is that?

Dr. Einstein: Because the fundamental theorem of arithmetic is this: Any positive integer (other than prime numbers), can be factored into primes in only one way, apart from the order in which the prime factors are written. There is no belief more "self-evident" to mathematicians. This is the Holy Grail. Be careful what you say.

JJA: Thanks for the warning. But I feel that you and I can talk freely about any topic. Am I wrong?

Dr. Einstein: No. No. Please continue.

JJA: First, I believe it was the German mathematician, Carl Jacobi (1804 -1851) who said, "One must always invert" an operation or procedure to discover other possibilities.

---

*For a simple explanation of primes, please see pages 11-13 and 11-14.

Dr. Einstein: That advice has paid off handsomely with many inventions and scientific discoveries.

JJA: Well, that is what I propose to do with primes. I want to "invert" and start with the notion that primes do not exist, and the reason no underlying pattern has been discovered is simply that primes do not exist.

Dr. Einstein: Something as basic to arithmetic as primes must have a hidden pattern. It is indeed baffling that no one in thousands of years has decoded this secret, not Pythagoras, not Euclid, not Fermat, not Descartes. Not even Newton.

JJA: Precisely. The reason is very simple. Primes do not exist. They are a mathematical illusion.

Dr. Einstein: What is your reasoning?

JJA: First, let's start with something very basic to arithmetic. Any arithmetic operation can only be performed in a pair. For example, we cannot add three numbers at once: $2 + 3 + 4 = ?$

What we actually do as a first step is add the 2 and the 3 to get 5, and then the second step is to add that 5 to the final 4... to get our answer, which is 9.

Dr. Einstein: And we cannot multiply three numbers at once. 2 times 3 times 4 = ?

What we actually do as a first step is multiply 2 times 3 to get 6, and then the second step is to multiply that 6 times the final 4... to get our answer, which is 24. Arithmetic works only in pairs, I agree. Now what?

JJA: Now, with the concept of pairs, let's look at a few numbers and apply the "fundamental theorem of arithmetic." For example, 8 = 2 times 4 which is a prime times a non-prime.

Dr. Einstein: But 8 also equals 2 times 2 times 2, all primes.

JJA: Not really. Let's break up the multiplication into pairs and examine the patterns. 8 = 2 times 2 = 4, and 4 times 2 = 8. Hence, we have a prime times a prime in the first pair and a non-prime times a prime in the second pair.

Dr. Einstein: Whooa! Let's try another example. How about the number 28 ?

JJA: The first pair is 2 times 7 = 14 and the second pair is 14 times 2 = 28. Again, we have a prime times a prime in the first pair and a non-prime times a prime in the second pair. Clearly, the factors are not all primes.

Dr. Einstein: So primes may not be basic to arithmetic! The implications are staggering. The search for the "Golden Equation" that will predict all primes is doomed to failure! No such equation exists.

JJA: My conjecture about primes is this: Where there is more than one pair in a series, only one of the pairs will have both factors as primes and at least one other pair will contain a non-prime.

An alternate model for arithmetic

Dr. Einstein: I liked your *copy and add* model of arithmetic which you presented to René Descartes. It was an elegant explanation for the fact that with fractions, the result increases when fractions are added but decreases when fractions are multiplied.

JJA: I thought you did not like the adjective "elegant" in mathematics. You once said that "elegance" is for tailors. Nature follows simple patterns.

Dr. Einstein: I did say that, but I think the adjective is appropriate for your *copy and add* model of arithmetic. Any other implications for the model?

JJA: Yes. I feel that the model can handle negative numbers, and especially the roots of negative numbers—as well as fractions.

Dr. Einstein: This is really exciting because the school model for arithmetic *cannot* find roots for negative numbers. For example, the square root of 9 is 3 but what is the square root of -9? If your model can find a root for -9, you have my full attention.

JJA: All right. First, let me remind you that the unifying principle of my model is that multiplication is repeated addition for negative and positive numbers—and fractions. The arithmetic model we learned in school, the school model you call it, seems to apply to positive integers only.

Dr. Einstein: Let's start with positive integers before we try negative integers. How would the Asher model handle the square root of positive integers such as the square root of 4?

JJA: Copy 2 twice and add like this: 2 + 2 = 4. Conclusion: the root of the square root of 4 is 2.

Dr. Einstein: Just to refresh my memory, in your *copy and add* model, the first "number" in a pair to be multiplied is not a number at all but a *mathematical instruction*.

JJA: Right! The first "number" in a pair to be multiplied instructs us what to do with the second number.

Dr. Einstein: Then the square root of 9 would be copy 3 thrice and add like this: 3 + 3 = 6 and 6 + 3 = 9.

Conclusion: The square root of 9 is 3. Now, how does your model find the square roots for negative numbers such as -4 or -9?

JJA: Let's try. For the square root of -4, use (2)(-2) which says to copy -2 twice and add like this: -2 + -2 = -4. The reverse is (-2)(2) which instructs us to copy 2 twice and reverse the result like this: 2 + 2 = 4 and the reverse is -4.

Conclusion: The square root of -4 is -2 or 2.

Dr. Einstein: Hmmm, you get two roots for the square root of -4.

Let me try the square root of -9. The pair would be (3)(-3) which says to copy -3 thrice and add like this: -3 + -3 = -6 and -6 + -3 = -9.

The reverse pair looks like this: (-3)(3) which says to copy 3 thrice and add, then reverse the results like this: 3 + 3 = 6 and 6 + 3 = 9 and the reverse is -9.

Conclusion: The square root of -9 is either -3 or 3. We get one root for the square root of positive integers and two roots for the square root of negative numbers. A curious result indeed! I wonder what this might represent in nature?

JJA: Can you explain what you mean?

Dr. Einstein: One of the great mysteries of the universe is that a mathematic which seems "empty" when it was discovered, later—in perhaps in the next generation or two turns out to represent some process in electricity or fluids or light. The hidden connection between numbers and natural events seems to me to be quite mystical.

JJA: Interesting! Any other suggestions for testing the validity of my model for arithmetic?

Dr. Einstein: How does the *copy and add* model apply to division?

JJA: Every arithmetic operation reduces to addition. Multiplication is transformed into addition in the *copy and add* model.

Dr. Einstein: And how about division?

JJA: Division is transformed into multiplication and then into addition. The basic or fundamental operation of arithmetic is addition.

For division, it works like this:

Step 1: 4 divided by 2 *or* 4/2 is transformed into multiplication like this: 1/2 times 4 *or* (1/2)(4).

Step 2: Multiplication is transformed into addition like this: Copy 1/2 of 4 which is 2 and add to zero to get 2.

Dr. Einstein: You are saying that there is really no such thing as multiplication or division, right? That is such a radical idea, I must try it myself. Let me start with 8 divided by 2, or 8/2.

For division, you say it works like this:

Step 1: 8 divided by 2 *or* 8/2 is transformed into multiplication like this: 1/2 times 8 *or* (1/2)(8).

Step 2: Then multiplication is transformed into addition like this: Copy 1/2 of 8 which is 4 and add to zero to get 4.

JJA: Yes. As usual, you are right on target.

Dr. Einstein: I like your model for three reasons: First, it's simplicity. We are dealing with only one arithmetic operation instead of four.* Second, the *copy and add* model explains many mysterious by-products in school arithmetic such as: Adding fractions gives an increase but multiplying fractions results in a decrease in value. Third, you demonstrate that indeed multiplication is simply repeated addition. I might add a fourth reason: The *copy and add* model gives roots for negative numbers.

I think school children learning arithmetic will have a more satisfying understanding of arithmetic with the *copy and add* model. That's the most exciting possibility.

---

*The fourth arithmetic operation is *subtraction* which is really only addition using negative numbers. To illustrate,

What we're used to seeing is this: 2 - 3 = -1
But the actual operation is this: 2 + (-3) = -1

Summary

There is no idea in mathematics more basic than the concept of prime numbers, the "building blocks" of all numbers. Primes are the DNA of mathematics. The mystery is this: If primes are so fundamental to numbers, why has the "golden equation" that will predict the location of primes in the series of natural numbers such as 1, 2, 3... still unknown?

Why has that basic of all equations eluded discovery by some of the most talented minds in history such as Pythagoras, Euclid, Fermat, Descartes, and Newton? Can it be that the answer is that "primes" are another illusion? Can it be that they do not exist? That is the premise that I explored with Dr. Albert Einstein.

Part II

Let's start off with a heated debate between the celebrated Pythagoras, and two amateur mathematicians, Farber and Heisel. Remember, the famous Pythagorean theorem has more "proofs" than any other idea in mathematical history. Yet Farber and Heisel don't believe the theorem is valid.

Yes, Pythagoras is certainly right when he proposed that for right-angle triangles, the length of the hypotenuse can be measured indirectly. First, find $a^2 + b^2 = c^2$ and then find $\sqrt{a^2 + b^2} = c$, where c is the hypotenuse. But Farber and Heisel argue that only in rare instances will $a^2 + b^2 = c^2$. Most often, c^2 is not a true square in the same sense that the following numbers are true squares: 4, 9, 16, 25, 36...

In a desperate attempt to retain the integrity of the revered Pythagorean theorem, Euclid searches for triplets, three numbers, that will fit into the equation of $a^2 + b^2 = c^2$ to produce true squares. In my conversation with Euclid, he reveals that his geometry book has been printed in over 2,000 editions, an unbelievable achievement in the world of publishing!

Next, the Prince of Amateur Mathematicians, Pierre de Fermat, reveals his "marvelously simple proof" to solve a mystery that has baffled mathematicians for hundreds of years. At one time in Europe, an award equal to several hundred thousand dollars was offered to anyone who solved this intriguing mystery. To this day, the award remains unclaimed.

The team of amateur mathematicians, Farber and Heisel, come back to take a fresh look at the nature of decimals which we all take for granted. They argue that the use of decimals is a serious mistake in calculation. Their reasoning appears to be most persuasive!

I engage in a conversation with Christian Goldbach about his famous Goldbach Conjectures, which have stymied mathematicians for 450 years. His explanation about the nature of odd and even numbers has a kindergarten simplicity and appears to be "obvious," but has yet to be proven. You may have an irresistible compulsion to try and solve this and if you succeed, you will be internationally famous overnight.

Finally, a favorite mystery of mine has to do with negative numbers. For centuries, mathematicians denied the existence of negative numbers. After all, negative numbers do not behave like positive numbers. You cannot square them, and you cannot find their square roots. They appeared to be imaginary entities. I invited the distinguished Augustine Louis Cauchy, to explore negative numbers with me and the outcome may surprise you.

My recommendation

Please relax and have fun with the ideas in Part II. You will see equations such as $(a + b)^2 = c^2$, which are the crown jewels of mathematics, but many people perceive them as a crown of thorns. Don't let the equations intimidate you. Remember, they are nothing more than simple arithmetic. Listen to Sir Isaac Newton whispering in your ear, "How difficult can it be?"

Best wishes for continued success,

James J. Asher

CHAPTER 15

A debate between the renowned Pythagoras and a team of amateur mathematicians, Farber and Heisel:

Is the world famous Pythagorean theorem a fiction?

Pythagoras: Gentlemen, you say that my theorem is a fiction even though I have demonstrated that in a right-angle triangle the square of each side equals a third square of the hypotenuse. In modern symbolism: $a^2 + b^2 = c^2$.

F-H: You demonstrated this relationship for $a = 3$, $b = 4$, and $c = 5$. This is hardly a universal proof.

Pythagoras: But others have found many triplets of whole numbers that fit the pattern of my theorem.

F-H: These are like a handful of sand in the Sahara Desert. Since there are so few instances, your theorem is non-representative. It remains to be proved. We notice that many of the triplets are merely multiples of $3^2 + 4^2$ such as $6^2 + 8^2 = 100$ and $9^2 + 12^2 = 225$.

Pythagoras: What do you suggest?

F-H: First, the sum of two whole numbers squared will not produce a third square of a whole number except in exceptional cases which are your triplets. It can't work because the third square is always the product of this pattern: $(a + b)^2$ which will work for all combinations of whole numbers. Notice that we underlined the word, "all."

Pythagoras: If you multiply the pattern of $(a + b)^2$, you get $a^2 + 2ab + b^2$. You say that the middle term of $2ab$ is essential.

F-H: Absolutely. The third square of a whole number is composed of three elements: $a^2 + 2ab + b^2$. Why you occasionally find whole numbers that fit a pattern of only $a^2 + b^2 = c^2$ is a puzzle yet to be deciphered, but the universal truth is that only $(a + b)^2$ will work for all whole numbers. Again, underline the word "all."

Pythagoras: Any examples of this?

F-H: Yes, indeed. Let's sample from the domain of whole numbers. Let's select a few whole numbers and compare your pattern with ours. Is that fair?

Pythagoras: Yes. First, let's try with $a = 1$ and $b = 2$...

Pythagorean Pattern
$a^2 + b^2 = c^2$
$1^2 + 2^2 =$
$1 + 4 = 5$
Conclusion: 5 is a non-square.

Amateurs' Pattern
$(a + b)^2 = c^2$
$(1 + 2)^2 =$
$1 + 4 + 4 = 9$ or 3^2
Conclusion: 9 is a true square.

F-H: Next, let's compare patterns with $a = 1$ and $b = 3$...

Pythagorean Pattern
$a^2 + b^2 = c^2$
$1^2 + 3^2 =$
$1 + 9 = 10$
Conclusion: 10 is a non-square.

Amateurs' Pattern
$(a + b)^2 = c^2$
$(1 + 3)^2 =$
$1 + 6 + 9 = 16$ or 4^2
Conclusion: 16 is a true square.

Pythagoras: Once more, let's try with **a = 2** and **b = 3**...

Pythagorean Pattern
$a^2 + b^2 = c^2$
$2^2 + 3^2 =$
$4 + 9 = 13$
Conclusion: 13 is a non-square.

Amateurs' Pattern
$(a + b)^2 = c^2$
$(2 + 3)^2 =$
$4 + 12 + 9 = 25$ *or* 5^2
Conclusion: 25 is a true square.

F-H: Our pattern of $(a + b)^2 = c^2$ works every time with whole numbers—*any **combination*** of whole numbers. This is a universal truth.

Pythagoras: What are the implications?

F-H: If one wishes to add the squares for any set of two whole numbers to produce a third square of a whole number, then <u>three</u> <u>elements</u> are necessary:

$$a^2 + 2ab + b^2$$

The **2ab** guarantees success every time. Without **2ab**, it is a rare chance that $a^2 + b^2$ will equal c^2 when **c** is a whole number.

Pythagoras: I notice something else about your pattern of $(a + b)^2 = c^2$.

F-H: What is that?

Pythagoras: With my pattern of $a^2 + b^2 = c^2$, I notice that c^2 will <u>sometimes</u> be a true square, but <u>most</u> <u>often</u> c^2 will turn out to be a non-square.

Now here is the rub. When I sit down at my computer and write a simple program, I can make my theorem work for every ***even*** integer and every ***odd*** integer.

F-H: This we gotta see. Without **2ab**, we don't see how this is possible.

Pythagoras: Well, let's explore. I want to start with odd integers such as 3, 5, 7 and so on.

F-H: Fine.

Pythagoras: I discovered that when **any odd** integer is squared (starting with the **odd** integer of **3**), I can run a computer program that I have written to find an appropriate **even** integer to complete the equation with a true square. In this example an even integer that will work is **4**.

Pythagorean Pattern for <u>Odd</u> Numbers such as 3, 5, 7, 9...
odd^2 + even^2 = true square
a^2 + b^2 = c^2
3^2 + 4^2 =
9 + 16 = 25 $or 5^2$
Conclusion: 25 is a true square.

Notice that we get the **5²** without the **2ab**. Perhaps you two can explain why. Here is a second example. I start with the next **odd** number which is **5**, and find an appropriate **even** number of **12**.

Pythagorean Pattern for <u>Odd</u> Numbers such as 3, 5, 7, 9...
odd^2 + even^2 = true square
a^2 + b^2 = c^2
5^2 + 12^2 =
25 + 144 = 169 $or 13^2$
Conclusion: 169 is a true square.

Now starting with the *odd* number of 7, an *even* number that will produce a true square is 24.

> **Pythagorean Pattern for <u>Odd</u> Numbers such as 3, 5, 7, 9...**
>
> $odd^2 + even^2 = $ true square
> $a^2 + b^2 = c^2$
> $7^2 + 24^2 = $
> $49 + 576 = 625$ or 25^2
>
> *Conclusion:* 625 is a true square.

Finally, starting with the *odd* number of 9, an *even* number that will produce a true square is 40.

> **Pythagorean Pattern for <u>Odd</u> Numbers such as 3, 5, 7, 9...**
>
> $odd^2 + even^2 = $ true square
> $a^2 + b^2 = c^2$
> $9^2 + 40^2 = $
> $81 + 1600 = 1681$ or 41^2
>
> *Conclusion:* 1681 is a true square.

Notice that the third square (on the right of the equals sign) is always the square of an *even integer* plus 1. That is, **c** equals **b + 1**. In the above example, **b = 40** and **c = b + 1** *or* **41**. This can also be proved with algebra.

F-H: Wow! We are impressed—but still unconvinced. But what happens when you <u>start</u> with *even* integers?

Pythagoras: All right. I find that for each *even* number squared starting with 2, 4, 6 and so forth, there is an *odd* number (a whole number and a fraction) squared, which, when added together will equal a third square. The third square will be a *whole number and a fraction* and, like the results when **a** is an odd integer, **c** will be **b + 1**. Let's start with the *even* integer of **2** and run the computer program I have written to find the appropriate *odd* number, which turns out to be $1\frac{1}{2}$. Take a look:

Pythagorean Pattern for <u>Even</u> Numbers such as 2, 4, 6, 8...
even² + odd² = square, but is it a *true* square?
a^2 + b^2 = c^2
$(2)^2$ + $(1\frac{1}{2})^2$ =
4 + $2\frac{1}{4}$ = $6\frac{1}{4}$ or $(2\frac{1}{2})^2$
Conclusion: $6\frac{1}{4}$ is a square, but not a *true* square.

Next, here is the pattern when I start with the *even* number of **4** and find the appropriate *odd* number, which in this case is $7\frac{1}{2}$:

Pythagorean Pattern for <u>Even</u> Numbers such as 2, 4, 6, 8...
even² + odd² = square, but is it a *true* square?
a^2 + b^2 = c^2
$(4)^2$ + $(7\frac{1}{2})^2$ =
16 + $56\frac{1}{4}$ = $72\frac{1}{4}$ or $(8\frac{1}{2})^2$
Conclusion: $72\frac{1}{4}$ is a square, but not a *true* square.

Now here is the pattern when I start with the *even* number of **6** and find the appropriate *odd* number, which is $17\frac{1}{2}$:

Pythagorean Pattern for <u>Even</u> Numbers such as 2, 4, 6, 8...
even² + odd² = square, but is it a *true* square?
$a^2 + b^2 = c^2$
$(6)^2 + (17\frac{1}{2})^2 =$
$36 + 306\frac{1}{4} = 342\frac{1}{4}$ or $(18\frac{1}{2})^2$
Conclusion: $342\frac{1}{4}$ is a square, but not a *true* square.

Again, notice that the *third* square (c^2) is always the square of **b + 1**, which can be shown with algebra

F-H: We applaud you, Maestro. That's a marvelous demonstration. The problem is that you do not have whole number solutions. Your theorem appears to work for odd numbers but not even numbers. Right?

Pythagoras: Yes. I am disappointed to say, yes.

F-H: Do you concede that our pattern will work for "all" whole numbers?

Pythagoras: Yes, you have convinced me.

F-H: And your pattern will work for odd numbers but not even numbers.

Pythagoras: That is correct.

F-H: Then, what are your findings telling us?

Pythagoras: There is quite a mystery here. I mean, trying to synchronize your findings with mine is rather perplexing. Let's review what we have: My discovery: There are two facts which seem quite clear.

First, if the square of "a" is odd, then the square of "b" will be even. The third square will be constructed from a number that is b + 1. Starting with an odd whole number, the solution will involve whole numbers for a, b and c.

F-H: But, ahhh, there is the "but"... you still agree that your pattern will not hold for any and all combinations of whole numbers.

Pythagoras: Yes, definitely.

F-H: It is interesting that given any <u>odd integer</u> you can find an <u>even integer</u> that will satisfy your pattern. But, given any <u>even integer</u>, only an <u>odd integer</u> and a <u>fraction</u> will produce a third square which will not be a whole number but rather a complex fraction.

Pythagoras: This is a puzzling turn of events. Clearly, you have the better pattern if we evaluate with a criteria of simplicity, harmony, and generality. But you will agree that my theorem produces some provocative results with odd numbers in "a."

F-H: Absolutely. But what does it all mean? There must be a bridge between our findings, but what is it? One more thought: Would you agree that while you have a theorem, we have a <u>law</u>?

Pythagoras: Oh, oh! Yes, yes. I would have to agree. My theorem can account for some integers but not all. Your law explains all combinations of integers. Also, my pattern is more complex compared with your simple pattern. We always prefer the simple to the complex. But, but... if I may point out, you gentlemen did not discover $(a + b)^2$. That formula for a square has been known for centuries.

F-H: Yes, but the connection to your theorem had not been clarified.

Pythagoras: I salute your effort. Indeed, you have pointed out some provocative implications.

F-H: We wonder what the great Euclid has to say about this controversy. As we recall, he discovered some "primitive" solutions for triplets.

Pythagoras: Primitive solutions for triplets? Can you explain?

F-H: These are whole numbers for a, b and c that are independent of other triplets.

Pythagoras: Well, it will be interesting to hear what Euclid has to say... But, if I may have the last word—my theorem does predict the length of the hypotenuse for any right-triangle. That seems firmly established.

F-H: Yes, indeed. Our debate concerns all those "proofs" that $a^2 + b^2$ will <u>always</u> equal c^2 when c^2 implies a true square of whole numbers such as 4, 16, 25...

Summary

Is the world famous Pythagorean theorem a fiction? Two amateur mathematicians think so and debate Pythagoras. The outcome is surprising.

The Pythagorean theorem has been proven in more ways that any other theorem in mathematics—it does measure the hypotenuse of a right-angle triangle, but does it result in "true squares"?

CHAPTER 16
Euclid explains why his book of geometry, reprinted in an unbelievable 2,000 editions, is still a best selling book after 24 centuries.

A Conversation with Euclid

JJA: Thank you for agreeing to this interview. Before we start may I ask you a few questions about your runaway bestseller, Elements?

Euclid: Certainly.

JJA: I checked out a copy from the library and was astonished. This has to be the most impressive work in the history of scholarly writing. For centuries, edition after edition of the Elements has been published.

Euclid: Yes, 2000 editions have been published at last count.

JJA: Incredible. How do you explain this phenomenal success?

Euclid: Of course, my contribution was simply to organize the work of the ancient mathematicians such as Archimedes.

JJA: Speaking of Archimedes, how do you feel about his famous quote: "Give me a place to stand and holding a lever, I can move the earth."

Euclid: Modern space exploration suggests that this may be possible under certain circumstances and zero gravity.

JJA: An enormous amount of work by the ancient mathematicians was destroyed when that splendid library at Alexandria was burned by the Roman Legions of Julius Caesar.

Euclid: Julius was a bastard. His mindless book burning frenzy inspired the Nazis to burn thousands of books in Germany circa 1939. Those Roman soldiers were butchers. They were models for Hitler's SS troops who also enjoyed senseless killing.

JJA: The Roman soldiers were the models for the SS?

Euclid: Absolutely. If a town resisted, once the Roman's occupied the town, they systematically slaughtered every living thing in the town—every man, woman, child and animal. They showed no mercy.

JJA: What a jolt! Let's relax and get back to things academic—in the research for your book, what did you discover about Archimedes? What sort of person was he?

Euclid: Completely absorbed in the search to decode the divine blueprint, the geometry of the creator. He etched diagrams everywhere—on paper, in the sand... and he would study these diagrams for days.

JJA: Diagrams. You and your predecessors did not have algebra as a tool.

Euclid: No. We did not know, for example, that 2 + 2 = 4 was an equation. It would have been thrilling for us to discover the concept of an equation.

JJA: Why?

Euclid: Because here was another tool the Creator must have used to form everything in nature. Equations are like a Chinese puzzle box. Once you open one equation, there is another hidden and then another. There are messages within messages about how the universe works.

JJA: It is truly amazing that you accomplished what you and your colleagues did with only a ruler, compass and drawing diagrams.

Euclid: Yes, thank you. Algebra would come centuries later with Arab scholars including the poet and mathematician, Omar Khayyám. Then a hundred years after my passing, along comes the Frenchman, René Descartes, with another stunning discovery: There is a code that will allow us to convert geometry to algebra and then decode from algebra to geometry. What a spectacular revelation, but why not talk with Descartes yourself about this? Be sure to ask him about his mystical experience with the "Angel of Truth."

JJA: I will, but for now let's get back to Archimedes, the man.

Euclid: Well, I shouldn't tell you this.

JJA: What?

Euclid: Archimedes literally stinks.

JJA: Whoa! You don't mean that.

Euclid: I do. He was so focused on his diagrams that he forgot about personal hygiene. His clothes, his body reeked with a foul odor. He was often carried high in the air by many men to a public bath.

JJA: Fascinating.

Euclid: Enough of other people. What else do you want to know about my book.

JJA: As I page through the Elements, I am stunned by the amount of work that must have been involved in producing this book.

Euclid: It wasn't work. It was play.

JJA: Play? How can it be play?

Euclid: You recall in your old testament, the story of ancient people who decided to trespass into God's residence by building a tower into the sky?

JJA: Right. the Tower of Babel. God thwarted their plan. They couldn't communicate with each other. Each artisan suddenly spoke a different language.

Euclid: To be face-to-face with God or at least discover traces of God's workmanship seems to be an eternal yearning. To the Greeks and others in the ancient world, revealing the secrets of geometry was exciting—actually thrilling because we were looking at the Creator's blueprint for the universe. We were decoding the work of the Divine Geometer. Your scholars experienced the same delirium with the discovery of the Dead Sea Scrolls hidden in desert caves for centuries. And there was the same unbearable excitement with the discovery of the Rosetta Stone. Remember?

JJA: I recall the reaction of one of our famous scientists, Albert Einstein, when someone gave him the gift of your book. He was 12 years old. He said that he opened your book and was electrified. It was the first thing he had ever experienced in his life where everything fit together with absolute perfection.

Euclid: Of course I cannot claim credit. I did not "invent" the relationships. They are products of the Supreme Mathematician. We merely discovered and deciphered the blueprint. But enough adulation. You wanted to talk about the Pythagorean theorem.

JJA: Yes. Well, you listened to the discussion between Pythagoras and the amateur mathematicians. I know that you discovered certain relationships between pairs of numbers that will produce triplets. What are your thoughts on this controversy?

Euclid: I myself found formulas for finding two squares that will add up to a third square. There are "primitive" solutions, meaning that each solution is independent of the others. None is merely a multiple of a previous triplet.

JJA: Ok. Now, for the first one hundred numbers squared, how many triplets do you find that are "primitive"?

Euclid: About 16.

JJA: About 16 in 100.

Euclid: We can almost double that number if we do not restrict ourselves to "primitives."

JJA: Then you conclude?

Euclid: Well, I have to conclude that my data do not support the Pythagorean theorem. At best, only 3 in 10 squares are triplets with whole numbers. Obviously, the Pythagorean theorem is non-representative. Only $(a + b)^2$ will generate a solution for every possible number squared.

JJA: So far the amateurs seem to be right on track.

Euclid: I noticed that Pythagoras used only one odd number to find triplets whereas I used a pair with one odd and the other even. How many triplets with whole numbers did he find for the square of numbers up to one hundred?

JJA: Only seven.

Euclid: Seven in one hundred. Well, again, his theorem is not supported by the evidence. Some, but not all square numbers can be decomposed into two other squares. Why this happens remains to be explained. To decompose a square into the sum of two other squares of whole numbers, only this pattern will work every time:
$a^2 + 2ab + b^2 = (a + b)^2$.

JJA: Wait! I have a sudden flash! Perhaps this is the answer: Every "real" number (which includes, integers, fractions, decimals, irrationals such as pi, etc.) can be squared.

Euclid: Right. Any number can be squared simply by multiplying the number by itself. We can even square an irrational such as pi.

JJA: Sure. Pi multiplied by pi is 1.4141... times 1.4141 = 1.9996... But notice that 1.9996... is not a true square.

Euclid: Every number can be squared to produce another number, but that number may not be a true square consisting of whole numbers.

JJA: Only certain numbers are true squares such as 4 is 2^2, 9 is 3^2, and 16 is 4^2.

Euclid: All other numbers are pseudo-squares. So, Pythagoras is right in his conclusion for any right-angle triangle which is half of a rectangle.

JJA: That conclusion being?

Euclid: That conclusion is that one side squared plus the adjoining side squared is equal to the hypotenuse squared, but...

JJA: But the hypotenuse squared is not necessarily a true square. For instance, $1^2 + 2^2 = 1 + 4 = 5$ (which is not a true square.

Euclid: For most right-angle triangles, the hypotenuse squared will not be a true square.

JJA: I think that the concept of true squares in the Pythagorean theorem clarifies this puzzle. What do you think?

Euclid: Definitely!

Summary

Euclid gets into the debate about the Pythagorean theorem. He also explains the runaway success of his geometry book which has been reprinted in an unbelievable 2,000 editions. He shares some gossip about the extraordinary Archimedes.

CHAPTER 17

A conversation with Pierre de Fermat, the Prince of Amateur Mathematicians.

Fermat reveals a *"marvelously simple demonstration,"* solving a mystery that has baffled mathematicians for hundreds of years.

Fermat's marvelously simple demonstration!

JJA: It is an honor to be interviewing the Prince of Amateur Mathematicians. During your time in the 1600s, your occupation was lawyer. Is that correct?

Fermat: Yes. My duty was to investigate charges of heresy among the clergy. If the charges were valid, then I had the task of prosecuting the violators.

JJA: Sounds a bit grim to my generation.

Fermat: It was a prestigious assignment in 15th century Europe. You must remember that the Church was the dominant institution in my time just as Economics is the most important institution in your time. You believe that free enterprise must be defended and preserved. We felt the same way about the Church and her doctrines.

JJA: Nice analogy. Tell me about your interest in mathematics. What earned you the title of Prince of Amateur Mathematicians?

Fermat: Prosecuting priests is stressful, as you can imagine. It was my duty and I did it, but it was not easy. To reduce stress, I played with mathematics.

JJA: You played with mathematics.

Fermat: Yes. I found that I had a natural talent and the more I dabbed, the better player I became.

JJA: You seemed to especially enjoy outwitting the professional English mathematicians.

Fermat: That was the most fun of all. They were sniffish about French mathematicians. They looked down their aristocratic noses at us. I enjoyed a cat and mouse game of toying with them, pulling their long, grotesque tails. Better get me off this subject.

JJA. All right. That cryptic note you left on the margin of the 1621 edition of Diophantus' classic book *Arithmetic* stirred up red-hot curiosity in the mathematical community since your passing in the 1600s. What was your "marvelously simple proof" all about? No one ever found any writing from you to follow up on this astonishing note.

Fermat: First, let me correct you. I did not mean that I had a marvelous *proof*. I intended to say that I had a marvelously simple *demonstration*—and you seem to have stumbled upon it.

JJA: How so?

Fermat: I listened to the debate between the amateur mathematicians and Pythagoras. Fascinating. The amateurs have unwittingly uncovered the solution, although they did not recognize it.

JJA: What is your demonstration all about?

Fermat: I can demonstrate that $a^3 + b^3$ *cannot* equal c^3. Next, $a^4 + b^4$ *cannot* equal c^4 and so forth.

JJA: The solution is in $(a + b)^2$?

Fermat: Yes. Yes.

JJA: Please explain.

Fermat: If you try $(a + b)^n$, you will generate c^n. The formula is perfectly general. It works for all powers from **2** to **n**.

JJA: For example?

Fermat: For example,
$(a + b)^3 = c^3$
$a^3 + 3(a^2)b + 3a(b^2) + b^2 = c^3$

JJA: Now let's see what happens when we plug in a few simple numbers.

Fermat: Sure. Try it.

JJA: All right. $(2 + 1)^3 = 27$ which has *cube root* of 3. Voilá! It worked.

Fermat: Exactly. Try it for $(2 + 1)^4$ and you get 81 which is the *fourth root* of 3.

JJA: And, $(2+ 1)^5 = 243$ which is the *fifth root* of 3. Zowie! Will this hold for all integers?

Fermat: Absolutely. All integers and all combinations of integers. This has the stature of a law, not just a theorem.

JJA: Now, how do you then conclude that there are no integers that will satisfy: $a^n + b^n = c^n$?

Fermat: It is obvious. From $(a + b)^n$, if you extract only a^n and add it to b^n you will not—absolutely not get c^n. For example, $a^n + b^n$ will not equal c^n.

JJA: Let's try it with the three examples above.

Fermat: Sure. For $(2 + 1)^3$, you get four terms:
8 + 12 + 6 + 1 which equals 27.
Extract the a^3 which equals 8.
Extract the b^3 which is 1.
Now add them together. What do you get?

JJA: $a^3 + b^3 =$
8 + 1 = 9

Fermat: And 9 has no cube root. Let's try the second example where $(2 + 1)^4$.

JJA: Okay. I get five terms of 16 + 32 + 24 + 8 + 1 = 81.

Fermat: Notice that $a^4 + b^4 =$
16 + 1 = 17

JJA: $2^4 + 1^4 = 17$, which does not have a root of 4.

Fermat: Now you're getting it. Shall we try one more just to really nail the mathematical coffin shut?

JJA: By all means. Let's consider $(2 + 1)^5$. There should be six terms and they are: 32 + 80 + 80 + 40 + 10 + 1.

Fermat: Correct. Now pull out a^5 which is 32 and b^5 which is 1. Now add them.

JJA: 32 plus 1 equals 33, which does not have root of 5.

Fermat: Want to continue?

JJA: I get it. But is this sufficient "proof" in the mathematical sense?

Fermat: If I promised a "proof," please let me off the mathematical hook with a "demonstration," which I believe is sufficient. In the genre of "proof," my demonstration is stronger than the demonstration of Pythagoras that resulted in his theorem—which, as you know, is riddled with holes. I honestly do not understand how his theorem was accepted as a universal truth for 24 centuries.

JJA: Any other suggestion for proof?

Fermat: Well, consider this analogy: H_2O = water. What kind of proof can one offer that this formula is true except to say: Try it. It works every time!

JJA: In other words, your demonstration holds up empirically since no contradictory examples have yet to be found in four centuries.

Fermat: Right. And none will ever be found.

JJA: Your conclusion again is this: $(a + b)^n = c^n$ but $a^n + b^n$ can never equal c^n for any power higher than 2.

Fermat: Yep. That is how the carriage wheel turns. Now I have a request.

JJA: By all means. What is it?

Fermat: Bring back the amateurs, Farber and Heisel, and let's explore decimals. I want to know why they call decimals "the greatest red herring in the history of mathematics."

Summary

Pierre de Fermat, the Prince of Amateur Mathematicians, reveals his "marvelously simple proof" for $a^n + b^n = c^n$. This has baffled mathematicians since the 15th century.

As a final note, Princeton Professor Andrew Wiles may finally have a proof for Fermat's Last Theorem. (For the complete story, see Simon Singh's book, *Fermat's Enigma*, which is listed under References in the back of this book.)

CHAPTER 18

A conversation about decimals with the amateur mathematicians, Farber and Heisel.

Are decimals real or just a mathematical illusion?

JJA: It never occurred to me to question decimals until I read your rare, out-of-print book published in 1931 called, *Mathematical and Geometrical Demonstrations*. But don't you think that your subtitle was a red flag, causing professional mathematicians to stampede—and trample your ideas!

F-H: Perhaps we were a bit too enthusiastic.

JJA: Enthusiastic? Mr. Heisel, you bill your work as inspired by a *"Brief and Infallible Method of Squaring the Circle Discovered by Mr. Theodore Faber of Brooklyn, N.Y..."* which was published in the Brooklyn, N.Y. Eagle, February 28, 1883.

Heisel: The word *'infallible'* was a mistake. Much too colorful.

JJA: That word is sure to infuriate the professionals who are always restrained. They are careful to monitor every word they write so that there are no unrealistic expectations. If anything, they understate, understate, understate...

F-H: Yes, we agree. They understate because they have good reason to understate.

JJA: Are you saying that they don't know anything?

F-H: No. Of course they are knowledgeable in their subject, but they have a point of view which they feel must be defended. They are prejudiced. It is like what happened to Dr. Edward Jenner, who suggested to the medical community in 1796 that he had a cure for the lethal small pox—a scourge that randomly executed men, women and children.

JJA: Dr. Jenner had a radical idea, as I recall.

F-H: Indeed. His solution was the ghastly idea to cut an "X" on the child's arm and smear the open wound with cow puss.

JJA: This was a daring and scary demonstration of what we now call "vaccination."

F-H: It was daring and scary—but this monumental contribution to medicine almost destroyed Dr. Jenner. The ridicule was relentless and unbearable. There were years of enduring abuse from physicians, but Dr. Jenner lived long enough to be vindicated. His vaccination worked when all other "cures" failed. After the year 1800, vaccination became an accepted medical practice.

JJA: Are you saying that your ideas have the same impact on society as Jenner's vaccination?

F-H: Not quite. But almost. Yes, our rhetoric was flamboyant. But you must remember that we are excited hobbyists who were thrilled with our mathematical discoveries. Neither of us has a university degree in the subject. So, please forgive us if we yell from the crowd, "Hey, the emperor has no clothes!" We report the truth as we see it.

JJA: I love it! But back to decimals.

F-H: Decimals are an illusion. Decimals are not a number but an operation.

JJA: How so?

F-H: First, it is important to note that the ancient Greek mathematicians did not use decimals in their calculations. As an aside, can you imagine trying to find decimals using Roman numerals?

JJA: I believe the concept of a decimal first appeared in India circa 600 A.D.

F-H: Correct. It wasn't until about the 1200s that the decimal system spread throughout Europe. Decimals became the new kids on the mathematical block.

JJA: Who introduced decimals into calculation?

F-H: Pythagoras must take some responsibility when he tinkered with right-angles and came up with his theorem that for any and all right-angled triangles, $a^2 + b^2 = c^2$. Notice he did not prove this was true. Quite the contrary, he selected a right-angle triangle having one side which has a length of 1 and another side with a length of 1. The hypotenuse of that triangle could <u>not</u> be the square root of 2 because multiplying any two *identical* integers or any two *identical* fractions together will not equal 2!

He was certain of this conclusion. This was the contradiction to $a^2 + b^2$ always equals c^2 in integers.

JJA: Then what happened?

F-H: Inexplicably, from then on, mathematicians accepted that the relationships between **a**, **b** and **c** was true, but the result was a novel and troublesome concept called an "irrational" number.

JJA: Intuitively, mathematicians must have realized that the equation was not universally true when they began a search for "triplets,"— that is, sets of three integers that satisfied the Pythagorean theorem of $a^2 + b^2 = c^2$.

F-H: Right.

JJA: Now what about these "troublesome, irrational numbers"?

F-H: It would have been better for Pythagoras if he never stumbled upon a few special cases where $a^2 + b^2 = c^2$ is true such as when $a^2 = 3$, $b^2 = 4$ and $c^2 = 5^2$. He was so intrigued with this finding that he was obsessed with experimenting using other integers— until he made the shocking discovery that is now world famous.

JJA: You mean his discovery that the relationship did not seem to work when the sides of the triangle were, for example, 1 and 1 (resulting in the hypotenuse being $\sqrt{2}$).

F-H: Yes. Now mathematicians must somehow explain, find or invent the $\sqrt{2}$ (which does not exist).

JJA: Hence, the introduction of the decimal.

F-H: Well, yes. When you try to take the square root of 2, you get a process of endless division. Out comes a stream of decimals like this 1.4142135... with no discernible pattern and the numbers seem to continue forever.

JJA: Mathematicians christened these new numbers "irrational" numbers.

F-H: And there is the illusion. The decimals are numbers but they are not numbers. This is a paradox that would have entertained Zeno and his colleagues for years.

JJA: If decimals are not numbers, what are they?

F-H: First, we should explore the concept of an "irrational" number.

JJA: All right.

F-H: Modern mathematicians classify numbers as either "rational" or "irrational." If a decimal can be converted into a fraction, it is rational. For example, the repeating decimal .333... can be converted into 1/3 and so it is rational. The repeating decimal .666... can be converted into 2/3, and so it too is rational. The non-repeating decimal .75 is rational because it converts into 3/4.

JJA: Most common fractions are rational.

F-H: Yes. You can convert a fraction into a decimal and then back into a fraction. Rational means ratio and a fraction is a ratio.

JJA: Got it. Now what? What makes some numbers "irrational"?

F-H: Irrational numbers cannot escape their decimal structure by a metamorphosis into fractions. For example, the square root of 2 is the endless decimal number 1.4142135... which can <u>not</u> be converted into a ratio. The square root of 3 is the endless decimal number 1.7320508... which also can <u>not</u> be converted into a ratio. By comparison, the square root of 4 is <u>rational</u> because it <u>can</u> be converted — into a ratio of 2/1.

JJA: So, where is the illusion?

F-H: The illusion is all those senseless numbers after the decimal point. For example, the square root of 2 looks like the discrete number 1.4142135... that is continually expanding forever with no repeating pattern. This does not make any sense.

A number by definition is discrete. It is definite. It is finite. A "number" <u>cannot</u> be changing, indefinite, and continuing into infinity. It cannot.

To understand this, we must delve deeper into the nature of decimals. Why do we have decimals at all? What do they mean? What is the "inner structure" of decimals?

JJA: We know how to generate decimals. We know how to do it? Do we know why they exist?

F-H: We think so. If we divide a fraction such as 1/2, we get a decimal of .50. 1/4 = .25 and 3/4 = .75. Intuitively, this makes sense. What does not make sense is 1/3 which equals .33333333... and so on with threes continuing forever.

Why didn't 1/3 terminate with .33? After all, 1/3 is a definite measurement on a ruler. It has a beginning and an end. It does not have a beginning, and an end plus a little bit more.

1/7 = .1428571... and continues on forever. This is puzzling when the measurement of 1/7 on a ruler has a beginning and a definite end. But the decimal equivalent has a beginning and no definite end.

JJA: 1/3 is .333... The pattern repeats itself. Does this make it definite?

F-H: No. The pattern repeats itself, but this tells us nothing except that the "wallpaper" pattern will go on and on forever to fill every surface in the universe and beyond. How come? There is a conflict between the measurement on a ruler and the conversion into a decimal. One result contradicts the other. This is a serious conflict. If we can figure this out, we will understand the nature of decimals.

JJA: Some fractions are no problem such as. 1/2 = .50, 1/4 = .25, 1/5 = .20 and 1/8 = .125 There is a clear-cut beginning and end to each fraction. There is harmony between the physical measurement and the decimals. Would you agree?

F-H: Yes. But, for other fractions such as 1/3, 1/6, 1/7, and 1/9 the results are completely contradictory. 1/3 = .33333..., 1/6 = .16666..., 1/7 = .1428571..., and 1/9 = .11111... When each fraction is converted into decimals, there is a beginning and <u>absolutely no end</u>.

JJA: I believe mathematicians created the concept of a "limit" to handle those troublesome fractions.

F-H: The concept of "limits" is an *ad hoc* attempt, a rather feeble attempt at that, to explain the paradox with the notion that "the fraction will approach some value which is called a "limit."

JJA: Do you have an alternate explanation?

F-H: The "inner structure" of the paradox becomes visible if we manually divide each of the troublesome fractions and convert into decimals like 1.0/3 = .3 What does the .3 mean?

JJA: It means 3 tenths.

F-H: What does 3 tenths mean?

JJA: It means that the point is located three units on a scale of ten.

F-H: You have a remainder of one when you divide 1/3. What will you do with the one?

JJA: Let's do what the teacher advises in grade school. Set the problem up like this 1.00/3 and we get .33

F-H: What does the .33 mean?

JJA: It means that on a scale of one hundred, the point is located 1/3 of the distance from the beginning of the scale. But we still have a remainder of one. So set the problem up again like this 1.000/3 and we get .333

F-H: Which means that your point is now located 1/3 of the distance on a scale of 1000 units. So, we can continue the division into eternity and what do you notice?

JJA: I notice that the scale is changing each time from a scale of ten to a scale of one hundred to a scale of one thousand.

F-H: Precisely. But what happens to the point you are trying to locate on the number line.

JJA: Nothing. It remains fixed at 1/3 of the distance from zero.

F-H: Ah, my dear Watson. You have penetrated the inner structure of the pattern.

JJA: When 1/3 is converted into decimals, each decimal that appears is merely a shift in the scale from 10 to 100 to 1000 and so forth. The value remains constant. No matter what the scale is, we are still one third of the distance on that scale.

F-H: And the red herring is that decimals give the illusion that the value is changing continually into infinity when in reality the value remains fixed and the scale is changing.

JJA: That's why you recommend that computation be limited to fractions.

F-H: Yes.

JJA: Whoa! That is such a radical idea.

F-H: If we take a closer look into the nature of division, you will appreciate the significance of operating only with fractions.

JJA: This should be interesting.

F-H: Let's start with a few examples: Ten divided by two equals five

JJA: Thanks for keeping it simple. I'm with you: 10/2 = 5. Now what?

F-H: The message here is that if we divide ten into two parts, each part will contain 5 numbers: 1, 2, 3, 4, 5 and 6, 7, 8, 9, 10.

JJA: So far, so good.

F-H: Now, another interpretation is this: 5 is a number located about in the <u>center</u> of a scale from 1 to 10.

JJA: Yes.

F-H: 100 divided by 2 equals 50 or in symbolic language: 100/2 = 50.

JJA: Yes.

F-H: This time the number is 50 and it is also located about in the <u>center</u> of a scale from 1 to 100.

JJA: I see where you are going. 1000/2 = 500 which is a number located about in the <u>center</u> of a scale from 1 to 1000.

F-H: Precisely. Conclusion: What do the numbers 5, 50, and 500 have in common? Taken in context of the scales, 5 and 50 and 500 are all are located about in the center of a scale. The values did not change, only the scales changed.

It is like looking out the window of a moving train. The scenery has the illusion of moving when in reality, the scenery is fixed while the train moves.

JJA: I am still bothered by your suggestion that computation be restricted to fractions. This could be cumbersome.

F-H: We believe that decimals are unnecessary in calculation. Decimals distract from exact measurement. Decimals give us approximations, at best, whereas calculation with fractions will give us precise measurements.

JJA: That is one I will have to think about. Using fractions in arithmetic operations of addition, subtraction, multiplication and division rather than decimals is like suggesting that we return to cumbersome Roman numerals such as I, II, III, IV to perform arithmetic rather than simple Arabic numbers of 1, 2, 3, 4.

F-H: In truth, it is more trouble to arithmetize with fractions than decimals. After all, calculators and computers are all programmed to calculate arithmetic with an output that is in decimals.

JJA: The ancients working only with integers were playing the mathematical golf game with the ball always on the green. When decimals were introduced, the mathematical golf ball was being whacked into sand traps.

F-H: For 24 centuries, mathematicians have been swinging at the ball trying to release it and send it soaring back onto the green. But all that happens is the whoosh of the club and a curtain of sand in the air. The ball is forever lost in the sand traps of infinity.

JJA: You're talking about the sand trap of "irrational" numbers.

F-H: Certainly. And also fractions converted into decimals that continue into infinity.

JJA: I believe I have an answer to that, but first, I would like to explore the concept of negative numbers with the greatest critic of negative numbers, Cauchy. As a final note, you are recommending to mathematicians a return to fractions in preference to decimals?

F-H: We know that this will be a "hard sell." This is like Galileo recommending to the medieval church that the earth is revolving around the sun when "everyone can verify with their own eyes by looking up that the sun is revolving around the earth." This will be a "hard sell," indeed. Remember that Galileo was almost executed until he recanted, on his knees and trembling with his head bowed, confessing: "I was in error. The sun is indeed revolving around the earth."

Summary

Are decimals real or just a mathematical illusion? Our two amateur mathematicians believe that decimals are the greatest "red herring" in the history of mathematics.

CHAPTER 19

A conversation with Christian Goldbach about his famous Goldbach Conjecture—an idea so simple that any elementary school child can understand it, yet it remains an unsolved mystery to this day.

JJA: Your name appears in the hundreds of books that explores number theory. Why do you think that is?

Goldbach: My idea is so simple; yet so illusive. It drives professional mathematicians crazy. It looks as if a kindergarten child can solve it.

JJA: Before we get into it, I was fascinated to discover that you knew Euler.*

Goldbach: Yes, although I was a German mathematician, Euler and I were both members of the Russian Academy of Science.

JJA: In 1725, you were also Professor of Mathematics in St. Petersburg, Russia.

Goldbach: And I had the honor of being a personal tutor to Tsar Peter II.

JJA: When did you propose the Goldbach Conjecture?

Goldbach: In a letter to Euler** on June 7, 1742.

---

*For a brief biographical sketch of Goldbach and Euler, please see *The Players* in the back section of this book.

**In Goldbach's letter to Euler, there was another intriguing bit of speculation: Every odd number greater than 7 is the sum of three primes. (The idea has yet to be proven or disproven.)

JJA: I am ready to take a crack at it. Can we talk?

Goldbach: By all means. My idea is this: I believe that every <u>even</u> <u>number</u> (other than 2) can be expressed as the <u>sum</u> <u>of</u> <u>two</u> <u>primes</u>. Such as:

4 = 2 + 2	14 = 3 + 11 and 7 + 7
6 = 3 + 3	16 = 3 + 13 and 5 + 11
8 = 3 + 5	18 = 5 + 13 and 7 + 11
10 = 3 + 7 and 5 + 5	20 = 3 + 17 and 7 + 13
12 = 5 + 7	22 = 3 + 19 and 5 + 17
	and 11 + 11

JJA: I understand exactly.

Goldbach: That was 250 years ago. As of now, my conjecture has been demonstrated to be correct up to about 10,000 even numbers.

JJA: Why is the Goldbach Conjecture about even numbers important?

Goldbach: Prime numbers are the DNA of mathematics because they explain how all the rest of the numbers are created. And since numbers are fundamental to mathematics, we want to know how primes work.

JJA: Is this a search for a "Divine" blueprint? Is this a concealed search for God's existence?

Goldbach: Perhaps... Yes. I would say, yes.

JJA: You are confident that your conjecture is correct up to 10,000 numbers, but the challenge is to prove it for all numbers.

Goldbach: Obviously, we cannot "prove" my conjecture by continuous expanding the test because it would go on without end.

JJA: Yes, but if someone finds just one counter example, this will <u>disprove</u> your conjecture.

Goldbach: Billions of examples will not prove my conjecture, and only one counter example will disprove it. That is the strange nature of mathematical thinking. The proof of my conjecture requires a logical argument which no one has yet developed.

JJA: Why do you think this has not yet happened?

Goldbach: The unknown factor is how to represent prime numbers such as 2, 3, 5... in <u>letters</u> such as **a, b, c**... Since the primes have no known pattern, we are not able to express them in <u>letters</u>.

JJA: Why is this important?

Goldbach: We are operating at the level of numbers such as 2, 3, 5, 7 and so forth. With letters we simplify the problem by moving to a higher level of abstraction. Mathematicians move up in abstraction to prove conjectures, while scientists move down from the abstract to concrete data to prove hypotheses. It is amazing to me that somehow these two worlds blend together in some mysterious way.

JJA: Please give me an example of a mathematical proof.

Goldbach: Well, let's take even and odd numbers. We have discovered many relationships showing how even and odd numbers behave because there is a pattern that we can express in letters.

JJA: Like what?

Goldbach: The <u>even</u> <u>numbers</u> can be represented by 2n and <u>odd</u> <u>numbers</u> by 2n + 1. For instance:

Table 1

2(n) = even number	2(n) + 1 = odd number
2(1) = 2	2(1) + 1 = 3
2(2) = 4	2(2) + 1 = 5
2(3) = 6	2(3) + 1 = 7

JJA: Let's see, with 2n, I can generate all the even numbers. If I add "1," we have all the odd numbers.

Goldbach: You got it!

JJA: Given that guidance, I am ready to try my hand at proving your famous conjecture.

Goldbach: Go for it!

JJA: The conjecture is: Every <u>even</u> <u>number</u> is the sum of <u>two</u> <u>prime</u> <u>numbers</u>. I would like to start with a simple idea and work my way up to your conjecture.

Goldbach: It is always wise to start with something simple.

JJA: I want to start with this idea: The sum of any two odd numbers will always equal an even number. For example:

Table 2

1 + 3 = 4	3 + 5 = 8	5 + 7 = 12
1 + 5 = 6	3 + 7 = 10	5 + 9 = 14
1 + 7 = 8	3 + 9 = 12	5 + 11 = 16

Goldbach: All right. Now what?

JJA: Now, using <u>letters</u> to represent <u>all</u> even and <u>all</u> odd numbers, I want to add two odd numbers to see whether the result is an even number. So, I add $(2n_1 + 1)$ to $(2n_1 + 1)$...

Goldbach: Good. What do you get?

JJA: $2n_1 + 1$
$\underline{2n_2 + 1}$
$2n_1 + 2n_2 + 2$ which can be reduced to: $2(n_1 + n_2 + 1)$

Goldbach: You get $2(n_1 + n_2 + 1)$. What do you make of it?

JJA: We know that 2 times any number will equal an even number because, as we showed in Table 1 on page 7-4:

$2(n)$ = even number
$2(1) = 2$
$2(2) = 4$
$2(3) = 6$

Hence, any number multiplied by 2 will be an even number, and the result of $2(n_1 + n_2 + 1)$ is an even number.

Goldbach: You have now "proved" that adding two odd numbers will always result in an even number. Bravo!

JJA: I like that word, "always." There are not many absolutes in the world, but here is one of them.

Goldbach: Where do you go from here?

JJA: Next, I know that odd numbers are composed of two types: primes and non-primes.

For instance, these <u>odd</u> <u>numbers</u> are ***primes***:
3, 5, 7, 11, 13, 17, 19, 23, and 29

But, these <u>odd</u> <u>numbers</u> are ***non-prime***:
9, 15, 21, 25, and 27

Goldbach: ...except for 2, <u>all</u> <u>the</u> <u>primes</u> are odd numbers.

JJA: But not all odd numbers are prime. Some are non-prime such as: 9, 15, 21, 25, and 27

Goldbach: Right.

JJA: Now, I have shown that the sum of two odd numbers will always equal an even number. Again, it works like this:

(odd number)   (odd number)

$(2n_1 + 1) \quad + \quad (2n_2 + 1) =$

$2(n_1 + n_2 + 1) \quad = \text{an even number}$

Since prime numbers (other than 2), are always odd numbers, then we conclude that the sum of two primes will always equal an even number. I believe this proves your famous Goldbach Conjecture.

Goldbach: Excellent show, but there is a fly in the mathematical ointment.

JJA: Oh, oh. What is it?

Goldbach: You have demonstrated, no stronger yet—you have "proved" that adding two primes will yield an even number.

JJA: That's not enough?

Goldbach: No. You must also show that the reverse is true. Any even number other than 2, can be decomposed into two primes.

JJA: Didn't I just show that?

Goldbach: We know that adding two odd numbers will always give us an even number.

JJA: The conclusion then is: An even number can be produced by adding two odd numbers—and these two odd numbers might be primes or non-primes.

Goldbach: Exactly.

JJA: I see why your conjecture is so maddening. We proved that the sum of two odd numbers will always produce an even number. We also proved that the sum of two primes (that are odd numbers), will produce an even number. Unfortunately, we did not prove that every even number is the sum of two primes (that are odd numbers).

Goldbach: We are so close, yet not quite there.

JJA: Can we explore the second intriguing bit of speculation in your letter to Euler?

Goldbach: You mean my conjecture about odd numbers?

JJA: Yes. Your idea, which has not yet been proven or disproven, is that <u>every</u> odd number greater than 7 is the sum of 3 primes.

Goldbach: Trying to prove that conjecture about odd numbers has driven mathematicians wild for the last 250 years.

JJA: Let me try to prove it.

Goldbach: Go!

JJA: Let's add 3 odd numbers like this:

$(2n_1 + 1) + (2n_2 + 1) + (2n_3 + 1) =$

$2n_1 + 2n_2 + 2n_3 + 3 =$

$2(n_1 + n_2 + n_3) + 3$

Goldbach: You get: $2(n_1 + n_2 + n_3) + 3$ Now, what does this show us?

JJA: It shows that the sum of 3 odd numbers equals an odd number! Since primes, except for 2, are all odd, can we not say that the sum of any set of 3 primes will be an odd number?

Goldbach: Yes.

JJA: Then your second conjecture about odd numbers is proved.

Goldbach: Again, not quite. We have the same problem that we had with even numbers. We have proved that the sum of three odd numbers produces an odd number. We have proved that the sum of three primes (that are odd numbers), produces an odd number. But what we have <u>not</u> proved is that: <u>every</u> odd number is the sum of three primes.

For example, we may find an odd number that is the sum of three odd numbers that are non-prime. I am thinking of 27, which is composed of the sum of three primes: 3 + 7 + 11, but it is <u>also</u> composed of three <u>non-primes</u>: 9 + 9 + 9. We may yet find an odd number composed exclusively of the sum of 3 non-primes.

JJA: So, the search goes on…

Summary

The mystery that has baffled mathematicians for 250 years is called Goldbach's Conjecture. His ideas have a kindergarten simplicity. First, every even number is the sum of two primes and the second is, every odd number is the sum of three primes. I think that I have possible proofs and I asked Goldbach to take a look.

CHAPTER 20

A conversation with Augustine Louis Cauchy.

Negative numbers - the greatest "red herring" in the history of mathematics?

JJA: You are famous for this quote in 1847: "(we discard) the symbolic sign square root of -1 which we repudiate completely and which we may abandon without regret, because one does not know what this alleged sign signifies, nor what meaning one should attribute to it."

Cauchy: I'm not the only doubter. There was Mahavira in India a thousand years before me who observed that in the nature of things a negative number has no square root and Cardan in 1545 in his *Ars Magna*, believed that negative numbers exist but the evidence is so doubtful that he concluded: negative numbers are fictitious. Any attempt to propose roots for negative numbers has a long mathematical history of denial.

JJA: Where do negative numbers come from?

Cauchy: We don't need them if we stick only to addition. It is when we introduce subtraction that negative numbers appear.

JJA: For example, if **a** is less than **b** and we subtract **b** from **a**, the result is a negative number. For instance, 5 - 10 = - 5.

Cauchy: Precisely. I have no problem with negative numbers on the number line. If you start at zero and go right, the integers are positive and if you go left from zero, the integers are negative. That seems simple, balanced, symmetrical —all attributes that we find in nature's blueprint. The problem is this: We can find roots for positive integers, so why not negative integers?

This does not fit into nature's design. For example, what is the square root of -1? There is no obvious answer. It is not (-1) times a (-1) because that equals a +1. The other problem is squaring negative numbers. It is jarring to the nervous system to have -1 times -1 = + 1. It does not make sense.

JJA: Standard school algebra gives a kind of fairy tale answer when we try to find the square root of negative numbers. For instance, let's consider the square root of -1.

It can't be 1, because 1 times 1 does not equal -1. It can't be -1, because in standard algebra, -1 times -1 equals +1. Hence mathematicians created an enigmatic "i" for a nonexistent $\sqrt{-1}$.

Cauchy: Then from that imaginary number called "i" (which, incidentally, I suggested), they created "complex" numbers

JJA: It sounds impressive.

Cauchy: I like Albert Einstein's view that nature obeys simple laws. If it is "complex," the model is probably not an accurate representation.

JJA: It is amazing how creative people have been in trying to justify negative numbers, especially multiplication with negative numbers.

Cauchy: I like what your 19th century essayist, Ogden Nash, had to say: "Minus times a minus results in a plus. The reason for this, we need not discuss."

JJA: One of the most creative interpretations that contorts all logic is this: The reason that a minus times a minus results in a plus is that when you multiply two positive numbers such as 2 times 2, you begin by standing on the number line at zero, turn to the right and walk forward four steps.

When you multiply -2 times a -2, you begin by standing at zero, turn left and then walk toe-to-heel so that you are moving backward in the direction of positive numbers. Walking backwards, you move four steps to +4!

Cauchy: The human mind has a rich imagination.

JJA: To be sure.

Cauchy: I noticed that your two amateur mathematicians challenged the reality of negative numbers such as the absurdity that -2 times a -2 = +4, but they did not pursue the issue.

JJA: Perhaps we can pick up where they left off. Schools give students the illusion that there is only one algebra.

Cauchy: Yes, most people do not know that in the history of mathematics, many algebras have been created.

JJA: You mentioned Einstein. To prove his theory of relativity, he used an algebra by the Irishman, Rowland Hamilton, to predict the location of Mercury.

Cauchy: ...which made Albert an international celebrity.

JJA: What does it take to create an algebra?

Cauchy: It is quite simple really. You need a few symbols and a few rules. If your system fits the rules with no contradictions, you have your own algebra.

JJA: Sound like it would be fun to try.

Cauchy: All right, let's try with the square root of -1. Can you think of a rule that gives us the square root of -1 without inventing a hideous concept such as "i" for imaginary number?

JJA: I think the simplest way to start is to use the standard symbols and rules of the algebra we in the 20th century were taught in school.

Cauchy: Good strategy. Now what?

JJA: I have played with this problem for some time and have concluded that it works best by reversing only one standard rule for multiplying signed numbers.

Cauchy: What rule did you reverse?

JJA: Only one...the one that states a negative times a negative equals a positive number.

Cauchy: So now you propose a negative times a negative will equal a negative number? I like it already because -2 times -2 will now equal -4.

JJA: But as I understand it, there are three properties of arithmetic that must be respected.

Cauchy: That's the three laws of multiplication: **Commutative, Associative**, and **Distributive**.

JJA: Right. Now if we reverse the standard rule for multiplying signed numbers, a negative times a negative is now negative.

Cauchy: -2 times -2 will now equal a -4.

JJA: Yes. But we may be in trouble with the Distributive Law which comes into play when we mix addition and multiplication.

The Distributive Law is written: **a(b + c) = ab + ac**. This means that we should get the <u>same</u> <u>answer</u> for **a(b + c)** as we get for **ab + ac**.

Let's see if this works with a few numbers, such as:
$a = -1, b = 2,$ and $c = 3$

Pattern #1	Pattern #2
a(b+c)	ab + ac
-1(2+3)	-1(2) + -1(3)
-1(5)	-2 + -3
-5	-5

Cauchy: Okay! It worked! You got a -5 for both patterns... what's wrong with that?

JJA: Nothing. But what if the numbers are:
$a = -1$, $b = -2$ and $c = 3$?

Pattern #1	Pattern #2
a(b+c)	ab + ac
-1(-2+3)	-1(-2)+ -1(3)
-1(1)	-2 + -3
-1	-5

Cauchy: Oh, oh! It didn't work with this example. One pattern gave us -1, and the other pattern gave us -5. This is a contradiction. The Distributive Law was violated! No matter how we play with it, it looks as if only (-2) times a (-2) will always equal +4. The mystery is why?

JJA: I have a possible explanation which I believe will give us three important advantages.

Cauchy: And they are?

JJA: First, we can eliminate the grotesque **"i"** for imaginary numbers along with the mutation called, "complex" numbers.

Cauchy: Sounds intriguing, if you can do it.

JJA: Secondly, my interpretation will allow negative numbers to have roots.

Cauchy: A striking achievement, if you can do it.

JJA: Finally, my interpretation will permit negative numbers to have squares.

Cauchy: Tell me how it works. I must know! Let me inspect the model for possible flaws.

JJA: The new interpretation for multiplying signed numbers is this: First, I remind you that in any arithmetic operation, we work with a pair only.

Cauchy: In other words, we cannot multiply three numbers or four. We are limited to only two numbers.

JJA: Exactly. Sounds obvious, but let's state it up front.

Cauchy: Now what?

JJA: Now, focus in on the sign of each number in a pair we want to multiply. For example, (-2) times (-2). We have assumed that we are multiplying two negative numbers. This, I believe, is an illusion, creating the mysterious (-2) times (-2) = +4.

Cauchy: You do not believe that we are multiplying two negative numbers?

JJA: No. I think it works like this: The first sign in the pair tells us to move to either the left or the right on a number line and the second sign tells us to continue in the same direction or reverse directions. Hence, if we multiply (-2) times (+2), the first sign which is negative tells us to move left on the number line and the second sign signals us to continue in the same direction.

Hence, (-2) times (+2) = -4.

Cauchy: Let's consider, (+2) times (-2). The first sign, according to your interpretation, tells us to move in the right direction on the number line and the second sign tells us to reverse direction. Hence, the result of (+2) times (-2) is a -4.

JJA: You got it! The first sign in a pair being multiplied tells us to move either to the right or to the left on the number line.

Cauchy: The second sign tells us to continue in the same direction or the reverse direction. I like it because there is a charming symmetry which is always pleasing in science or mathematics. Now what?

JJA: With my interpretation We have a new option. We can now find the square root for negative numbers. For example, $\sqrt{-4} = (+2)(-2)$ or $(-2)(+2) = -2$
$$\sqrt{-9} = (+3)(-3) \text{ or } (-3)(+3) = -3$$
$$\sqrt{-16} = (+4)(-4) \text{ or } (-4)(+4) = -4$$
$$\sqrt{-25} = (+5)(-5) \text{ or } (-5)(+5) = -5$$

Cauchy" Let's see...What is the essence of square root? It means that we find two identical numbers which when multiplied together will equal the number under the radical.

JJA: The illusion has been that the sign of each number in the pair represents a direction either right or left on the number line when actually only the first sign represents direction.

Cauchy: The second sign signals us to either continue in the same direction or reverse direction. So, each sign in the pair means something different.

JJA: And therefore, the signs should not be considered an integral part of the number. Multiply absolute numbers and use the signs for direction only.

Cauchy: I like it! It is a fresh look at signed numbers. So we can now find the square root of negative numbers which eliminates the need for the imaginary "i". How about the squares of negative numbers?

JJA: Let's try it: -4 = (+2)(-2) or (-2)(+2).

Cauchy: You need a new symbol for the relationship.

JJA: How about this: - 4 = [±2]
- 9 = [±3]
-16 = [±4]
-25 = [±5]

Cauchy: Bravo! What happens to poor (-2)(-2)? Why is it equal to +4?

JJA: I think that this is a false mystery created when we imagined that we were multiplying two negative numbers. We wondered why if multiplication is simply repeated addition—why don't we get a negative number when we multiply two negatives.

Cauchy: The answer is: We were NOT multiplying two negative numbers because the first sign signals direction—left or right and the second sign tells us to continue in the same direction or reverse direction. A number is only negative in isolation as, for instance, -2 which indicates that we are located two steps to the left of zero on the number line.

JJA: But, in addition or subtraction, we are indeed dealing with "pure" positive or negative numbers. The sign in each member of a pair tells us that we are moving either left or right on the number line. For example, (+2) + (+2) tells to move two steps to the right of zero on the number line and move right again to a +4.

(+2) + (-2) tells us to move two steps to the right of zero on the number line, then turn and move two steps to the left which leaves us back at zero.

Cauchy: I think that works.

JJA: Also, I notice that (+2) times (+2) = +4, and (-2) times (-2) = +4 look asymmetrical, but they show a charming symmetry when we graph it like this:

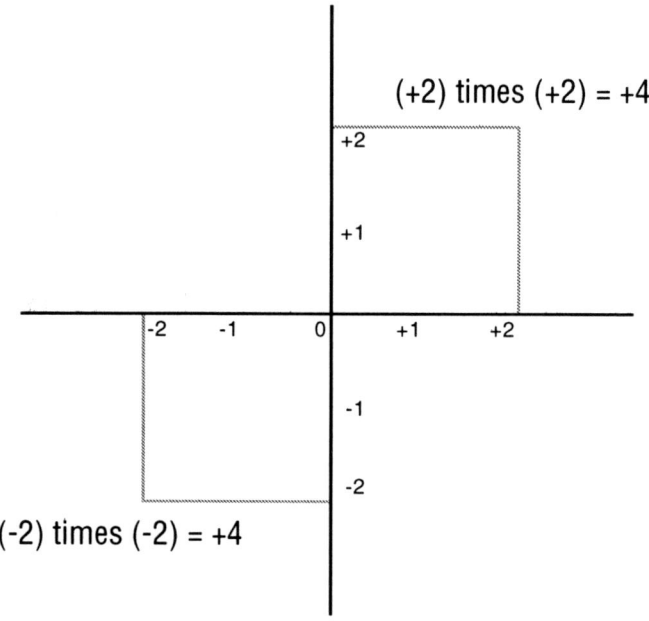

Summary

Standard algebra that we learned in school results in a mystery: 2 times 2 = +4 but—and the mystery is this: Why is -2 times -2 not equal to -4 ? If multiplication is simply repeated addition, then a symmetrical relationship should look like this:

    (+2) + (+2) = +4   (+2) times (+2) = +4
    (-2) + (-2) = -4   (-2) times (-2) = -4

So why is that (-2) times (-2) = +4? The exciting answer presented in this chapter is this insight: When we *multiply* two signed numbers, there is the false assumption that for each number a positive sign indicates the number is located to the right of zero on the number line and a negative sign tells that the number is located to the left of zero on the number line.

For example, if we *multiply* (+2) times a (-2), the assumption is that we are dealing with a 2 located two steps to the right of zero and another 2 located two steps to the left of zero.

A better interpretation is this: For (+2) times a (-2) multiply 2 times 2 to get 4. Now, the first sign, which is positive, tells that we are moving to the right on the number line and the second sign, which is negative, signals us to reverse direction. Hence, we wind up four steps to the left of zero.

When multiplying a pair of signed numbers, the first sign indicates direction, either left or right on the number line and the second sign tells us either to continue in the same direction or reverse direction.

The benefits of the new interpretation: First, there is symmetry with the first sign telling us to turn either left or right and the second sign telling us to continue in the same direction or reverse direction.

Next, the new interpretation allows us to find, for the first time in the history of mathematics, the square and the square root of negative numbers which Cauchy calls "an astonishing achievement" because we can vanish the grotesque invention of the imaginary **"i"** and the resulting mutations called "complex" numbers. This suggests that addition, subtraction, and multiplication are now closed with integers. Previously, arithmetic was not closed with integers because, for example, $\sqrt{-4}$ resulted in the enigmatic 2i. Now the $\sqrt{-4} = [\pm 2]$.

The Players in
Mathematical Mysteries

Archimedes (287 - 212 B.C.)

Often called "one of the three greatest mathematicians of all time" and the "father of experimental science." He invented the law of the lever and the pulley—his play toys that inspired the construction of machines able to move heavy loads easily. His famous comment: "Give me a place to stand, and I will move the earth."

He was the Jules Verne of his time, inventing ingenious ways to defend Syracuse (a Greek colony) from Roman assault. The weapons: A catapult that hurled huge rocks at the enemy, grappling hooks that could seize ships and disable them, and a system of mirrors that focused the sun's rays to set enemy ships on fire.

He introduced the concept of "limits" for the area of a circle—an idea that inspired the development of calculus. He showed that the value of **pi** is between 3 1/7 and 3 10/71—a discovery that made it possible to solve many problems involving the area of circles and the volume of cylinders.

Cauchy, Augustine-Louis (1789 -1857)

Frenchman who captured the concept of "limits" in a precise mathematical fashion. Trained as a military engineer, he was persuaded by Joseph Louis La Grange to leave engineering and devote himself to mathematics and pure science. Established modern calculus based upon a set of theorems with rigorous definitions. His publications in mathematics comprise 26 volumes and a journal for his mathematical writing alone.

Descartes, René (1596 - 1650)

Because of frail health as a child born into a wealthy family, he was allowed to stay in bed as long as he wished, a habit that remained for the rest of his life. In the warmth of his comfortable bed for hours, he would think about mathematical mysteries.

In 1616, he earned a degree in law but decided upon the military. Although he experienced combat, he was a gentleman soldier who had ample time to play with philosophical and mathematical ideas. His stunning breakthrough was the 100 page book on the scientific method that demonstrated the mysterious connection between geometry and algebra—a connection that mathematicians for centuries believed did not exist. This is known as analytic geometry and was the foundation for Newton's discovery of calculus.

Einstein, Albert (1879 -1955)

The caricature of him is that of a mild-mannered, absent- minded, white-haired, eccentric scientist scribbling esoteric calculations on a chalkboard. This image has become an icon for the public perception of professors everywhere. But Dr. Einstein, winner of the Nobel prize for physics, was a gentle, attractive man who was fascinated with politics, religion, and philosophy. He had musical talent, was widely read, and interested in all aspects of life.

As to his childhood, when Einstein's father asked the head master of the school, "What *occupation do you think my son is best suited?*" The reply: *"It doesn't matter what he does. He will not be successful at anything."* Einstein, later in life, was fond of saying that, *"Imagination is more important than intelligence."*

Euclid (about 300 B. C.)

Three hundred years before the birth of Christ, a Greek mathematician by the name of Euclid wrote a book about geometry that is still the best selling book in history behind the Bible. Starting with a few axioms (assumptions), Euclid deduced a complete system of Geometry. Others have used Euclid as a model such as Peano in the 19th century, who created five axioms to produce arithmetic.

When Dr. Einstein was twelve years old, he received as a gift, a copy of Euclid's geometry. Opening the book, Einstein felt the hair on the back of his neck stand up because he saw, for the first time, relationships that seemed to fit together with perfection. However, critics would point out many riddles in Euclid's work that to this day are unsolved such as *Proposition 42,* in which Euclid describes a parallelogram that shall be equal to a given triangle. How can a parallelogram with four sides be the same (equal) to a three-sided triangle?

Euler, Leonhard (1707 -1783)

For his eulogy, 50 pages were required to list titles of his published work which included some 700 books and papers. He had an uncanny ability to calculate in his head. He is credited with introducing "π" for **pi**, "*e*" to represent the base of natural logarithms, and "*i*" for $\sqrt{-1}$. Losing the sight in one eye, he was sometimes called the "mathematical Cyclops."

Farber, Theodore (circa 1800)

Farber was an amateur mathematician who had the audacity to wrestle not one, not two but three respected gorillas that have roamed the mathematical jungle for centuries. One is the Pythagorean theorem, the second is the value of decimals, and the third is the validity of negative numbers.

Fermat, Pierre de (1601 - 1665)

Sometimes called the *Prince of Amateur Mathematicians,* Fermat founded modern number theory. He was a French lawyer specializing in the prosecution of clergy suspected of heresy. For relaxation from his stressful occupation, he played with mathematical mysteries. His famous conundrum that has baffled mathematicians for hundreds of years is a cryptic note scribbled in the margin of a book: *"I have discovered a marvelously simple proof that this margin is too narrow to contain."* This is known worldwide as *Fermat's Last Theorem.*

Goldbach, Christian (1690 - 1764)

Professor of Mathematics in St. Petersburg, Russian, Goldbach was a member of the Russian Academy of Science and personal tutor to Tsar Peter II. The famous *Goldbach Conjecture* has stumped thousands of mathematicians for the past 250 years. In a letter to Euler on June 7, 1742, he speculated that every even number greater than 2 is the sum of two primes and every odd number greater than 7 is the sum of three primes.

Heisel, Theodore (circa 1930)

Heisel was an amateur mathematician who wrote a book based upon the audacious contention that the Pythagorean theorem, celebrated for 24 centuries as a universal truth in geometry—that this "truth" is a fiction. He was so convinced that he and Farber had discovered some "universal truths" that professional mathematicians had overlooked, that he self-published five thousand copies of his book and distributed them free to public libraries throughout the United States of America.

Newton, Isaac (1642 - 1727)

The *"greatest mathematician that England ever produced"* entered Cambridge at nineteen with no skill in mathematics beyond arithmetic; yet he created a stunning new branch of mathematics that could measure "curved" lines. He discovered the true nature of white light and the laws of universal gravitation.

Pythagoras (580 -500 B. C.)

He believed that number was the "cause" of everything in nature. Formed a secret society where "knowledge is the greatest purification" and knowledge meant mathematics. He was fascinated with the concept of simple ratios of whole numbers that generated the most beautiful musical melodies and celestial harmony. He is perhaps the most often quoted mathematician in history. After three years of listening to the master's voice from behind a curtain, students graduated from his famous school—a school that continued to function for 500 years. There is no student in school today who has not heard of the celebrated Pythagorean theorem.

Asher, James J.

Dr. Asher is an applied research statistician who discovered the *Mathematical Rosetta Stone*—a way to find a "Pythagorean" ratio for every integer before and after the decimal point in the so-called "irrational" numbers. He is **"JJA"** in this book, the interviewer of other players exploring mathematical mysteries.

Invitation to the reader:
Dr. Asher welcomes your comments and suggestions. You can contact him by e-mail at **tprworld@aol.com**

SELECTED REFERENCES

Asher, James J. *Brainswitching: Learning on the Right Side of the Brain.* 2001. Sky Oaks Productions, Inc., P.O. Box 1102, Los Gatos, CA 95031 (To order online, click on **www.tpr-world.com**)

Asher, James J. *The Super School: Teaching on the Right Side of the Brain.* 1999. Sky Oaks Productions, Inc., P.O. Box 1102, Los Gatos, CA 95031 (To order online, click on **www.tpr-world.com**)

Asher, James J. *Learning Algebra on the Right Side of the Brain.* 2001. You can read this article on the web by clicking on **www.tpr-world.com**

Beckmann, Petr. *A History of Pi.* 1971. St. Martin's Press, 175 Fifth Ave, New York, NY 10010

Burton, David M. *History of Mathematics: An Introduction.* 1995. Wm. C. Brown Communications, Inc., 2460 Kerper Blvd., Dubuque, IA 52001

Buxton, Laurie. *Mathematics for Everyone.* 1985. Schocken Books, New York, NY

Clawson, Calvin C. *Mathematical Mysteries: The Beauty and Magic of Numbers.* 1996. Plenum Press, 233 Spring Street, New York, NY 10013-1578

Devlin, Keith. *Mathematics: The Science of Patterns.* 1994. W. H. Freeman and Co., 41 Madison Avenue, New York, NY 10010

Esty, Warren W. *The Language of Mathematics.* 1995. Order from: www.math.montana.edu/~umsfwest/index.html/ or from Warren W. Esty, 511 Henderson, Bozeman, MT 59715

Feynman, Richard P. *Surely You're Joking, Mr. Feynman!* 1989. W. W. Norton and Co., 500 Fifth Ave., New York, NY 10110

Feynman, Richard P. *What Do You Care What Other People Think?* 1988. W. W. Norton and Co., 500 Fifth Ave., New York, NY 10110

Greene, Brian. *The Elegant Universe: Superstrings, Hidden Dimensions, and the Quest for the Ultimate Theory.* 1999. W. W. Norton and Company, 500 Fifth Ave., New York, NY 10110

Guillen, Michael. *Bridges to Infinity: the Human Side of Mathematics.* 1983. Jeremy P. Tarcher, Inc., 9110 Sunset Blvd., Los Angeles, CA 90069.

Hersh, Reuben. *What is Mathematics Really?* 1997. Oxford University Press, Inc., 198 Madison Ave., New York, NY 10016

Lederman, Leon and Teresi, Dick. *The God Particle: If the Universe is the Answer, What is the Question?* 1993. Dell Publishing, 1540 Broadway, New York, NY 10036

Meschkowski, Herbert. *Ways of Thought of Great Mathematicians: An Approach to the History of Mathematics.* 1964. Holden-Day, Inc., 728 Montgomery Street, San Francisco, CA

Pappas, Theoni. *The Joy of Mathematics: Discovering Mathematics All Around You.* 1989. Wide World Publishing, P.O. Box 476, San Carlos, CA 94070.

Pappas, Theoni. *Mathematicsl Scandals.* 1997. Wide World Publishing, P.O. Box 476, San Carlos, CA 94070.

Rees, Martin. *Just Six Numbers: The Deep Forces That Shape the Universe.* 1998. Doubleday, 1540 Broadway, New york, NY 10036

Singh, Simon. *Fermat's Enigma.* 2000. Basic Books, 10 E. 53rd Street, New York, NY 10022-5299

Tobias, Sheila. *Overcoming Math Anxiety.* 1978. Houghton Mifflin Company, 2 Park Street, Boston, MA 02108

White, Michael and Gribbin, John. *Einstein: A Life in Science.* 1993. Penguin Books, Inc., 375 Hudson Street, New York, NY 10014

White, Michael. *Isaac Newton: The Last Sorcerer.* 1997. Addison-Westley, Reading, MA

TPR PRODUCTS

**Books • Games
Student Kits
Teacher Kits
Audio Cassettes
Video Demonstrations**

Order directly from the publisher using your
<u>VISA</u>, <u>MASTERCARD</u>, *or* <u>DISCOVER CARD</u>
WE SHIP ASAP TO ANYWHERE IN THE WORLD!

Sky Oaks Productions, Inc.
TPR World Headquarters Since 1973
P.O. Box 1102
Los Gatos, CA 95031 USA

Phone: (408) 395-7600
Fax: (408) 395-8440
E-mail: tprworld@aol.com

FREE CATALOG UPON REQUEST!

For up-to-the-minute TPR Updates
and upcoming TPR Workshops, click on:

www.tpr-world.com

Latest books by James J. Asher
Originator of the Total Physical Response, known worldwide as TPR

Our Best Seller!

✓ **Demonstrates** step-by-step **how to apply TPR** to help children and adults acquire another language **without stress.**

✓ More than **150 hours** of classroom-tested TPR lessons that **can be adapted to teaching any language** including Arabic, Chinese, English, French, German, Hebrew, Spanish, Japanese, and Russian.

✓ A behind-the-scenes look at how **TPR** was developed.

✓ **Answers more than 100 frequently asked questions** about **TPR**.

✓ **Easy to understand** summary of 25 years of research with Dr. Asher's world famous **Total Physical Response**.

Learning Another Language Through Actions
by
James J. Asher
Originator of the
World Famous
Total Physical Response
Over 50,000 Copies in Print!

Order #201

NEW FEATURES
- Frequently Asked Questions - Newly Expanded!
- Letters from my mailbag
- e-mail addresses for TPR instructors around the world

Sky Oaks Productions, Inc.
P.O. Box 1102 • Los Gatos, CA, USA 95031
Phone: (408) 395-7600
Fax: (408) 395-8440
e-mail: tprworld@aol.com

Free TPR Catalog
upon Request!
or order directly online from
www.tpr-world.com

Latest books by James J. Asher
Originator of the Total Physical Response,
known worldwide as TPR

Brainswitching:
Learning on the Right Side of the Brain

2nd Ed. - 308 Pages   For *Fast, Stress-Free Access* to   Order #202
Languages, Mathematics, Science, and much, much more!

The Super School:
Teaching on the Right Side of the Brain
*To help most students learn anything fast
in academics, sports, or technology!*   Order #204
Your students won't want to miss a single class!

NEW!
The Weird and Wonderful World of
Mathematical Mysteries
Conversations with famous scientists and mathematicians.
by James J. Asher   Order #91a

Exclusive - new discovery published here for the first time solves a 2,000 year old mystery that baffled such famous people as Pythagoras, Euclid, Sir Isaac Newton, and Einstein.

- Dr. Asher successfully removed the fear of learning a foreign languages with TPR. Now, he removes the fear of mathematics!
- Conversations with famous scientists and mathematicians reveals their secret strategy for making spectacular breakthroughs by playing with ideas on the right side of the brain.
- Dr. Asher demonstrates how anyone with the basic skills of addition, subtraction, multiplication, and division can have a shot at world fame by finding hidden patterns in nature.

A Simplified Guide to Statistics
New! for Teachers, Principals, and Administrators
by James J. Asher

✓ Shows how to make a powerful presentation using data rather than anecdotes—to win the enthusiastic support your innovations deserve from policy-makers!
✓ Best tool in your instructional toolbox to show others what really works.
✓ Loaded with practical examples with straightforward explanations in plain English.
✓ Asher demonstrates step-by-step how to gather and analyze data to support your ideas for improving language instruction.   Order #266
✓ College students struggling with statistics will love this book too!

*Added **Bonus**:* Also learn how to organize a successful master's thesis or doctoral dissertation.

Dear Colleague:

Language instructors often say to me, "I tried the TPR lessons in your book and my students responded with great enthusiasm, but what can the students do **at their seats?**"

Here are effective TPR activities that students can perform **at their seats.** Each student has a kit in full color, such as the interior of a kitchen. Then you say in the target language, "Put the man in front of the sink." With your kit displayed so that it is clearly visible to the students, you place the man in the kitchen of your kit and your students follow by performing the same action in their kits.

As items are internalized, you can gradually discontinue the modeling. Eventually, you will utter a direction and the students will quickly respond without being shown what to do.

Each figure in the **TPR Student Kits** will stick to any location on the playboard **without glue.** Just press and the figure is on. It can be peeled off instantly and placed in a different location over and over.

You can create fresh sentences that give students practice in understanding hundreds of useful vocabulary items and grammatical structures. Also, students quickly acquire "function" words such as **up, down, on, off, under, over, next to, in front of,** and **behind.**

To guide you step-by-step I have written ten complete lessons for each kit (giving you about 200 commands for each kit design), and those lessons are now available in your choice of **English, Spanish, French,** or **German.** The kits can be used with **children or adults** who are learning **any language** including **ESL** and the **sign language of the deaf.**

About the TPR Teacher Kits

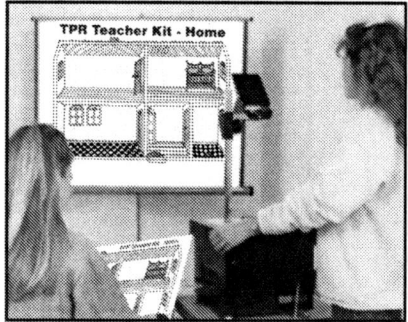

Use the **transparencies** with an overhead projector to flash a playboard on a large screen. Your students **listen** to you utter a direction in the target language, **watch** you perform the action on the large screen, and then follow by performing the same action in their **Student Kits.**

Best wishes for continued success,

James J. Asher

P.S. My sister and I recently tried one of the Student Kits with a native speaker of Arabic giving directions. We were both surprised at how much vocabulary and grammar we picked up in only a few minutes of play.

Back By Popular Demand!

Your favorite kits are back in production! Buy 6 Kits (Student or Teacher) in any assortment, and select an additional Kit as our **Free Gift** to you!

James J. Asher's TPR STUDENT KITS™

More than 300,000 Kits now being used in FL-ESL classes throughout the world!!

	ENGLISH Order Number	SPANISH Order Number	FRENCH Order Number	GERMAN Order Number
Airport ©	4E	4S	4F	4G
Beach ©	12E	12S	12F	12G
Classroom ©	10E	10S	10F	10G
Department Store ©	13E	13S	13F	13G
Farm ©	60E	60S	60F	60G
Garden ©	17E	17S	17F	17G
Gas Station ©	5E	5S	5F	5G
Home ©	1E	1S	1F	1G
Hospital ©	21E	21S	21F	21G
Kitchen ©	2E	2S	2F	2G
Main Street ©	15E	15S	15F	15G
NEW ➤ Office ©	6E	6S	6F	6G

(Includes high tech business machines such as computers, cell phones, fax, satellite communications, and more!)

Picnic ©	16E	16S	16F	16G
Playground ©	20E	20S	20F	20G
Restaurant ©	40E	40S	40F	40G
Supermarket ©	11E	11S	11F	11G
Town ©	3E	3S	3F	3G
NEW ➤ European Map ©	23E	23S	23F	23G

(Recently updated to include the Middle East!)

United States Map ©	22E	22S	22F	n/a
4-KITS-IN-ONE ©	50E	50S	50F	50
Calendar © (limited supply)	31	(In English)		

James J. Asher's TPR TEACHER KITS™
Transparencies for an <u>Overhead</u> <u>Projector</u>

	ENGLISH Order Number	SPANISH Order Number	FRENCH Order Number	GERMAN Order Number
Airport ©	4ET	4ST	4FT	4GT
Beach ©	12ET	12ST	12FT	12GT
Classroom ©	10ET	10ST	10FT	10GT
Department Store ©	13ET	13ST	13FT	13GT
Farm ©	60ET	60ST	60FT	60GT
Garden ©	17ET	17ST	17FT	17GT
Home ©	1ET	1ST	1FT	1GT
Hospital ©	21ET	21ST	21FT	21GT
Kitchen ©	2ET	2ST	2FT	2GT
Main Street ©	15ET	15ST	15FT	15GT
NEW ➤ Office ©	6ET	6ST	6FT	6GT

(includes high tech business machines such as computers, cell phones, fax, satellite communications, and more!)

Picnic ©	16ET	16ST	16FT	16GT
Playground ©	20ET	20ST	20FT	20GT
Supermarket ©	11ET	11ST	11FT	11GT
Town ©	3ET	3ST	3FT	3GT
NEW ➤ European Map ©	23ET	23ST	23FT	23GT

(includes the Middle East!)

United States Map ©	22ET	22ST	22FT	n/a

For over 25 years, Ramiro Garcia has successfully applied the Total Physical Response in his high school and adult language classes.

This Triple-expanded Fourth Edition (over 300 pages) includes:
- ✓ Speaking, Reading, and Writing
- ✓ How to Create Your Own TPR Lessons.

And more than 200 TPR scenarios for beginning and advanced students.
- ✓ TPR Testing for all skills including oral proficiency.
- ✓ TPR Games for all age groups.

Instructor's Notebook:
How to Apply TPR For Best Results
by
RAMIRO GARCIA
Recipient of the
OUTSTANDING TEACHER AWARD
Edited by
James J. Asher

Order #225

In this illustrated book, Ramiro shares the tips and tricks that he has discovered in using TPR with hundreds of students. No matter what language you teach, including ESL and the sign language of the deaf, you will enjoy this insightful and humorous book.

New! Just off the press! THE SEQUEL!!!

Instructor's Notebook: TPR Homework Exercises
by
RAMIRO GARCIA
Recipient of the
MOST REMEMBERED TEACHER AWARD
Edited by
James J. Asher

Ramiro's brand-new companion book to the Instructor's Notebook!
- ✓ Hundreds of TPR exercises your students can enjoy at home
- ✓ Catch-up exercises for students who have missed one or more classes.
- ✓ Review of the classroom TPR experience at home
- ✓ Helps other members of the student's family to acquire another language.
- ✓ Helps the teacher acquire the language of the students with exciting self-instructional exercises!

Order #224

Sky Oaks Productions, Inc.
P.O. Box 1102 · Los Gatos, CA, USA 95031
Phone: (408) 395-7600
Fax: (408) 395-8440
e-mail: tprworld@aol.com

Free TPR Catalog upon Request!
or order directly online from
www.tpr-world.com

The Graphics Book©

 For Students of All Ages acquiring
English, Spanish, French, or German
by
RAMIRO GARCIA

Dear Colleague;
You recall that I introduced graphics in the **Instructor's Notebook**. Hundreds of teachers discovered that **students of all ages** thoroughly enjoyed the material.

Your students understand a huge chunk of the target language because you used TPR. Now, with my new *graphics* book, you can follow up with **300 drawings** on tear-out strips that help your students *zoom ahead* with **more vocabulary, grammar, talking, reading** and **writing** in the target language.

In this book, you will receive **step-by-step guidance** in how to apply the *graphics* effectively with **children and adults** acquiring **any** language including **ESL**.

As an **extra bonus**, I provide you with **60 multiple-choice graphic tests for beginning and intermediate students.** Order **The Graphics Book** in your choice of English, Spanish, French or German.

TPR BINGO©
by Ramiro Garcia

In 25 years of applying the **Total Physical Response** in my high school and adult Spanish classes, **TPR Bingo** is the one game that students ask to play over and over!

When playing the game, students hear the instructor utter directions in the target language. As they advance in understanding, individual students ask to play the role of caller, which gives them valuable practice in **reading and speaking.** For an extra bonus, students **internalize numbers** in the target language from 1 through 100.

TPR Bingo comes with complete step-by-step directions for playing the game and rules

Use your Visa or Mastercard to order from anywhere in the world • We ship ASAP!

for winning. There are 40 playboards (one side for beginners and the reverse side for advanced students). A master caller's board is included, with 100 pictures, chips, and caller-cards in your choice of English, Spanish, French or German. As I tell my colleagues, "Try this game with your students. You will love it—they will love it!"

Brand-new feature! *Now included in every order of TPR Bingo...*
Play TPR Bingo with your students to move them from the imperative to the declarative (and interroga
It's easy, it's fun, and you will love it!

How to TPR Vocabulary!

- Giant 300 page resource book, alphabetized for quick look-up.
- Yes, includes *abstractions!*
- Yes, you will discover how to TPR 2,000 vocabulary items from Level 1 and Level 2 textbooks.

Where
Look up the word...   **For all ages!**
How to TPR it
1. Pedro, stand up and run to the door. Maria, sit **where** Pedro was sitting. 2. Write the name of the country **where** you were born. 3. Touch a student who's from a country **where** the people speak Spanish (French, English).

The **Command Book**
How to TPR 2,000 Vocabulary Items in Any Language
by STEPHEN SILVERS

Order #273

How to TPR Grammar!
For Beginning, Intermediate, and Advanced Students of All Ages!
Available for English, Spanish, and French!

"TPR is fine for commands, but how do I use it with other grammatical features?"
Eric Schessler shows you how to apply **TPR** for
stress-free acquisition of 50 grammatical features such as:

Abstract Nouns	Expletives	Object Pronouns	Possessive Pronouns	Simple Present
Adjectives	Future - to be going to	Past Continuous	Prepositions of Place	Singular/Plural
Adverbs	Future - Will	Past Perfect	Prepositions of Time	Nouns
Articles	Have - Present and Past	Past tense ofBe	Present Continuous	Subject Pronouns
Conjunctions	Interrogative Verb forms	Possessive Case	Present Perfect	Tag Questions
Demonstratives	Manipulatives	and Of expressions	Simple Past	Verbs
				Wh - Questions

English Grammar Through Actions   (Order #260)
Spanish Grammar Through Actions   (Order #261)
French Grammar Through Actions   (Order #262)

FAVORITE GAMES
FOR
FL - ESL CLASSES   Order #291

(For All Levels and All Languages)
by
Laura Ayala & Dr. Margaret Woodruff-Wieding

Chapter 1: Introduction

Chapter 2: Getting Started with Games
- How to get students involved
- How the games were selected or invented.

Chapter 3: Game Learning Categories
- Alphabet and Spelling
- Changing Case
- Changing Tense
- Changing Voice
- Describing Actions
- Describing Objects

Chapter 3 (Cont.)
- Getting Acquainted
- Giving Commands
- Hearing and Pronouncing
- Statements & Questions
- Negating Sentences
- Numbers and Counting
- Parts of the Body and Grooming
- Plurals and Telling How Many
- Possessive Adjectives & Belonging
- Recognizing Related Words
- Telling Time
- Using Correct Word Order.

Chapter 4: Games by Technique
- Responding to Commands
- Guessing   • Associating
- Simulating   • Sequencing
- Listing   • Matching
- Categorizing

Chapter 5: Special Materials For Games
- Objects   • Cards
- Authentic Props • Stories
- Pictures

Chapter 6: Bibliography

Actionlogues
By Jody Klopp Jody Klopp

More LIVE ACTION
in English, Spanish, French, or German!

✔ 25 happenings come to life in 396 photographs!
Examples: Getting ready for work, making a sandwich, going on a date, driving a car, etc.
✔ Internalize 160 verbs.

✔ Native speaker on a cassette utters each direction in the target language. Listen and understand instantly by looking at a photo.
✔ **Added Bonus:** Great way for non-native language teachers to expand vocabulary.

TPR IS MORE THAN COMMANDS —AT ALL LEVELS
CONTEE SEELY & ELIZABETH ROMIJN

Order #95

Winners of the
EXCELLENCE IN TEACHING AWARD
Explodes myths about the Total Physical Response:
Myth 1: TPR is limited to commands.
Myth 2: TPR is only useful at the beginning.

Demonstrates how to use James Asher's approach—

✔ to *overcome problems* in the use of TPR,

✔ to teach *tenses* and *verb forms* 6 ways,

✔ to teach *grammar, idioms,* and *fluent discourse,*

✔ to help your students *tell stories* that move them into fluent speaking, reading, and writing.

✔ Shows you how to go from zero to correct spoken fluency with TPR.

Very practical, with many examples!

Winner of the Paul Pimsleur Award!

Detailed lesson plans for **60 hours** of TPR Instruction that make it **easy** for novice instructors to apply the total physical response approach **at any level.** The lessons include:
- **Step-by-step directions** so that instructors **in any language** *(including ESL)* can apply comprehension training successfully.
- **Competency tests** after the 10th and 30th lessons.
- **Pretested short exercises** - that capture student interest.
- **Many photographs**

NOTE!
We have printed the lessons in two languages — **English** and **German**, but we have charged you only the cost of printing a single language.

COMPREHENSION BASED LANGUAGE LESSONS
LEVEL 1

By
Margaret Woodruff, Ph.D.
Winner of the Paul Pimsleur Award
(With Dr. Janet King Swaffar)

Order #290

TPR for Young Children!

Marvelously simple format: Glance at a page and instantly move your students in a logical series of actions.

- **Initial screening test** reveals each student's skill.
- After each lesson, there is a **competency test**.
- Recommended for beginning students in **preschool, kindergarden,** and **elementary.**
- Order in **English (#240), Spanish (#241),** or **French (#242).**

TPR for Children of All Ages!

For 20 years, "Listen & Perform" worked for children of all ages learning English in the Amazon - and it will work for your students too!

TPR Student Book by STEPHEN MARK SILVERS
Edited by **James J. Asher**

Your students in elementary and middle school will enjoy more than 150 exciting pages of stimulating right brain **Total Physical Response** exercises such as: *drawing • pointing • touching • matching • moving people, places, and things*

With the **Student Book** and **Cassette,** each of your students can perform <u>alone at their desks</u> or <u>at home</u>. TPR Lessons for students in elementary and middle school. Chock-full of fun and productive TPR activities for <u>older students</u> too!

Order in <u>English</u>, <u>Spanish</u>, or <u>French</u>!

TOTAL PHYSICAL RESPONSE IN THE FIRST YEAR

By
Dr. FRANCISCO L. CABELLO
with **William Denevan**

Dear Colleague:

I want to share with you the **TPR Lessons** that my high school and college students have **thoroughly** **enjoyed** and **retained** for weeks—even months later. My book has...

- A step-by-step script with props for each class.
- A command format that students thoroughly enjoy. (Students show their understanding of the spoken language by successfully carrying out the commands given to them by the instructor. **Production** is delayed until students are ready.)
- Grammar taught implicitly through the imperative.
- Tests for an evaluation of student achievement.

Sincerely,

Francisco Cabello, Ph.D.

Hot off the press in your choice of
English (#221), Spanish (#220), or French (#222)!

Look, I Can Talk!

High School, College, or Adults!

Student Book for Level 1
in English, Spanish, French, or German
by Blaine Ray
with Greg Rowe and Greg Buchan

Here is an effective **TPR** storytelling technique that **zooms** your students into *talking, reading,* and *writing*. It works beautifully with beginning, intermediate and yes — even advanced students.

Step-by-step, Blaine Ray shows you how to tell a story with **physical actions**, then have your students *tell the story to each other* in their own words **using the target language**, then **act** it out, **write** it and **read** it.

Each **Student Book for Level 1** comes in your choice of *English, Spanish, French* or *German* and has

✔ 12 main stories

✔ 24 additional action-packed picture stories

✔ Many options for retelling each story

✔ Reading and writing exercises galore.

Blaine **personally guarantees** that each of your students will eagerly tell stories in the target language by using the **Student Book**.

To insure rapid student success, follow the thirteen magic steps explained in the **Teacher's Guidebook** and then work with your students story-by-story with the easy-to-use **Overhead Transparencies**.

Look, I Can Talk More!

The Exciting Sequel!

Student Book for Level 2
in English, Spanish, French, or German
by Blaine Ray
with Joe Neilson, Dave Cline, Carole Stevens, and Christopher Taleck

For updates on current TPRS Workshops in your area, click on www.tpr-world.com

Keep the excitement going with this sequel for your level 2 students. Ten main stories with many spin-off mini-stories for variety. The drawings are superb and **Overhead Transparencies** may also be ordered. Your students will love it!

Look, I'm Still Talking!

Now...

Student Book for Level 3
in English, Spanish, French, or German

Fluency Thru TPR Storytelling

(Tells you how to do it - step by step!)   Order #96

Yes, demonstration videos are also availabe!

New! TPR Storytelling
especially for students in elementary and middle school
by
Todd McKay

- ✔ Pre-tested in the classroom for 8 years to guarantee success for your students.
- ✔ Easy to follow, step-by-step guidance each day for three school years - one year at a time.
- ✔ Todd shows you how to switch from activity to activity to keep the novelty alive for your students day after day.
- ✔ Evidence shows the approach works: Kids in storytelling class outperformed kids in the traditional class.
- ✔ Each story comes illustrated with snazzy cartoons that appeal to young people.
- ✔ There is continuity to the story line because the stories revolve around one family.
- ✔ Complete with tests to assess comprehension, speaking, reading and writing.
- ✔ Yes, cultural topics are included.
- ✔ Yes, stories include most of the content you will find in traditional textbooks, including vocabulary and grammar.
- ✔ Yes, included is a brief refresher of classic TPR, by the originator— Dr. James J. Asher.
- ✔ Yes, games are included.
- ✔ Yes, your students will have the long-term retention you expect from TPR instructions.
- ✔ Yes, Todd includes his e-mail address to answer your questions if you get stuck along the way.

TPR Products by Todd McKay:
- **Illustrated student booklet** for: Year 1, Year 2, or Year 2. *(In English, Spanish, or French.)*
- **Teacher's Guidebook** *(In English, Spanish, or French.)*
- **Complete Testing Packet** for listening, speaking, reading, and writing.
- **Transparencies**
- **Video Demonstration** shows you how to perform successfully step-by-step.

TPR Index Cards - NEW!
1. 4x5 index cards tell you exactly what to say, lesson by lesson
2. 60 Cards with First Year vocabulary.
3. No need to fumble through a book
4. No need to make up your own lessons
5. Quick! Easy to use! Classroom-tested for success!
6. Works for students of all ages

TPR Index Cards available for:
English: Order #470   **French**: Order #472
Spanish: Order #471   **German**: Order #473

For the latest updates about Todd McKay workshops, click on www.tpr-world.com

Best Demonstrations of Classic TPR *Anywhere in the World!*

James J. Asher's Classic Videos demonstrate the original research...
Historic videos show the original TPR research by Dr. James J. Asher with children and adults learning Japanese, Spanish, French and German. These vintage demonstrations are time-tested, and even more valid today than when the film was shot decades ago. We include with every video cassette a copy of the scientific publications documenting the amazing results you will see. A must for anyone teaching TPR. Each video cassette is unique, and shows different stress-free features of TPR instruction — *no matter what language you are teaching*, including English as a Second Language. *(Each video is narrated in English. Remastered from the original 16mm films.)*

Rent or Buy! Children Learning Another Language: An Innovative Approach©

VHS, Color, 26 minutes, shows the excitement of children from K through 6th grades as they acquire **Spanish** and **French** with **TPR**. (ESL students will enjoy this too!)

If you are searching for ways that motivate children to learn another language, don't miss this classic video demonstration. The ideas you will see can be applied in your classroom for any grade level and for any language, including English as a second language.

Rent or Buy! A Motivational Strategy for Language Learning©

VHS, Color, 25 minutes, demonstrates step-by-step how to apply **TPR** for best results with students between the ages of 17 and 60 acquiring **Spanish**. Easy to see how **TPR** can be used to teach any target language.

See the excitement on the faces of students as they understand everything the instructor is saying in Spanish. After several weeks in which the students are silent, but responding rapidly to commands in Spanish, students spontaneously begin to talk. You will see the amazing transition from understanding to speaking, reading, and writing!

Rent or Buy! Strategy for Second Language Learning©

VHS, Color, 19 minutes, shows students from 17 to 60 acquiring **German** with **TPR**. Applies to _any_ language!

Even when the class meets only two nights a week and no homework is required, the retention of spoken German is remarkable. You will be impressed by the graceful transition from understanding to speaking, reading, and writing!

Rent or Buy! Demonstration of a New Strategy in Language Learning©

VHS, B&W, 15 minutes, shows American children acquiring **Japanese** with **TPR**. Applies to _any_ language! You will see the first demonstration of the **Total Physical Response** ever recorded on film when American children rapidly internalize a complex sample of Japanese. You will also see the astonishing retention one year later! Narrated by the Originator of TPR, Dr. James J. Asher.

CPSIA information can be obtained at www.ICGtesting.com
Printed in the USA
BVOW041609210213

313822BV00001B/5/A